THE WINES OF ROUSSILLON

THE CLASSIC WINE LIBRARY

There is something uniquely satisfying about a good wine book, preferably read with a glass of the said wine in hand. The Classic Wine Library is a series of wine books written by authors who are both knowledgeable and passionate about their subject. Each title in The Classic Wine Library covers a wine region, country or type and together the books are designed to form a comprehensive guide to the world of wine as well as an enjoyable read, appealing to wine professionals, wine lovers, tourists, armchair travellers and wine trade students alike.

Other titles in the series include:
Port and the Douro, Richard Mayson
Sherry, Julian Jeffs
Madeira: The islands and their wines, Richard Mayson
The wines of Austria, Stephen Brook
The story of champagne, Nicholas Faith
The wines of Faugères, Rosemary George MW
Côte d'Or: The wines and winemakers of the heart of Burgundy, Raymond Blake
Rosé: Understanding the pink wine revolution, Elizabeth Gabay MW
Amarone and the fine wines of Verona, Michael Garner
The wines of Greece, Konstantinos Lazarakis MW
Wines of the Languedoc, Rosemary George MW
The wines of northern Spain, Sarah Jane Evans MW
The wines of New Zealand, Rebecca Gibb MW
The wines of Bulgaria, Romania and Moldova, Caroline Gilby MW
Sake and the wines of Japan, Anthony Rose
The wines of Great Britain, Stephen Skelton MW
The wines of Chablis and the Grand Auxerrois, Rosemary George MW
The wines of Germany, Anne Krebiehl MW
The wines of Georgia, Lisa Granik MW
The wines of South Africa, Jim Clarke
The wines of Portugal, Richard Mayson
Wines of the Rhône, Matt Walls
The wines of Piemonte, David Way

THE WINES OF ROUSSILLON

ROSEMARY GEORGE MW

ACADEMIE DU VIN LIBRARY

Rosemary George MW was lured into the wine trade by a glass of The Wine Society's Champagne at a job interview and subsequently became one of the first women to become a Master of Wine (1979). The author of thirteen books, she has been a freelance wine writer since 1981. Her very first book, *Chablis and the Wines of the Yonne*, published in 1984, won both the André Simon and the Glenfiddich awards; a second edition followed in 2007. For The Classic Wine Library she has written *The wines of Chablis and the Grand Auxerrois, Wines of the Languedoc* and *The wines of Faugères*. She contributes to various magazines, such as *Decanter* and *Sommelier India*, and writes a regular blog, www.taste-languedoc.blogspot.com. She is the current President of the Circle of Wine Writers.

First published in 2021 by Infinite Ideas Limited
This edition published 2024 by Académie du Vin Library Ltd
academieduvinlibrary.com

A CIP catalogue record for this book is available from the British Library.

ISBN 978–1–913141–81–3

Front cover © CIVR
Back cover: © Emilio Pérez
Photos: plate 1 (top), plate 2, plate 3, plate 4 (top), plate 5 (bottom), plate 6 (bottom), plate 8 © CIVR; page 60, plate 1 (bottom) courtesy of Mas Amiel; plate 4 (bottom), plate 7 (bottom) © Emilio Pérez; page 47, plate 5 (top) © Daiga Zilgave; plate 6 (top) Yvan Schreck; page 40, plate 7 (top) © Pierre Parcé.

Maps created by Darren Lingard from originals supplied courtesy of CIVR.

Printed in the U.S.A.

CONTENTS

INTRODUCTION: SETTING THE SCENE

Roussillon stands alone, proud and independent. For many years it suffered a union of convenience with the Languedoc, when the wines of two relatively unknown areas lacked any reputation, and when it was simpler to refer to the departments of the south, without differentiating between them, as Languedoc-Roussillon. Roussillon deserves so much more than that; it needs to come out from under the shadow of the Languedoc and stand alone. Its history is different, its language is different and the wines are quite different and original. Much of Roussillon is Catalan, with strong links to Catalonia in Spain. The local language is Catalan, whereas the Languedoc is part of Occitanie, where Occitan is the local language. Roussillon did not become fully part of France until the Treaty of the Pyrenees in 1659.

The original reputation of Roussillon is founded on what are rather clumsily called Vins Doux Naturels, VDN for short, fortified wines for which the key grape varieties are Muscat and Grenache, with their easily attained high alcohol levels making them the most suitable varieties. The natural sweetness is the sugar that remains in the juice after fortification. The wine growers tend to simply refer to the Vins Doux Naturels as vins doux, in contrast to their table wines, or vins secs, but I have preferred to use their full name, to emphasize their originality and how they differ from unfortified sweet wine or vin doux. Vins secs, as the unfortified table wines are commonly called, are a relatively recent development in Roussillon. It is only in the last 20 years or so that vins secs have overtaken Vins Doux Naturels in importance. The first appellations in Roussillon appeared in 1936 and were for Vin Doux Naturel; the first appellations for vins secs did not follow until 1971, for red Collioure, and 1977 for

Côtes du Roussillon and Côtes du Roussillon Villages, admittedly a few years before the key appellations of the Languedoc, in 1985.

Essentially Roussillon equates to the department of the Pyrénées-Orientales. Its boundaries are limited by the Pyrenees, with the Canigou the highest peak, at 2,785 metres, providing an important landmark. To the north, the foothills of the Corbières massif separate it from the Languedoc vineyards of Corbières itself, with the ruined Cathar castle of Quéribus and the lookout tower of Tautavel dominating the skyline.

Usually I have approached Roussillon from the Languedoc, driving south on the motorway. You pass Fitou, the last village of the Languedoc, reaching the extraordinary fortress of Salses shortly after, by which time you are in Roussillon and in northern Catalonia. In the distance is the outline of the Pyrenees, which are snow-capped for much of the year. These mountains unify the two halves of Catalonia. The fortress of Salses was first constructed in the fifteenth century and adapted by Vauban in the seventeenth century. It is well worth a visit, which you can easily do from the motorway *aire*, without leaving the motorway itself. Taking a slower and more scenic route over the hills from Corbières, you come past the castle of Quéribus and descend into the Agly Valley past either Vingrau and Tautavel or Maury. Alternatively, for still more dramatic scenery, there is the most stupendous of all the Cathar castles, the Château de Peyrepertuse, from which you could take in the Gorges de Galamus to St Paul-de-Fenouillet.

Three principal rivers cross the region to meet the Mediterranean. The most northern is the Agly, with the twin appellations of Maury and Maury Sec, as well as many of the villages of Côtes du Roussillon Villages. The Agly Valley really is the core of the vineyards of Roussillon, with awe-inspiringly majestic scenery. The vineyards peter out after Caudiès-de-Fenouillèdes, as the climate becomes cooler. If you carry on west along the valley past Axat and drive through the dramatic Defile de Pierre-Lys, with its threatening overhanging rocks, the next vineyards you encounter are those of fresher, more bucolic Limoux. The middle river is the Têt, which flows past the city of Perpignan and the northern edge of Les Aspres, where the vineyards are on undulating slopes. Then to the south there is the Tech, which meets the sea just north of the resort of Argelès and the vineyards of Banyuls and Collioure. Even without wine, Collioure and Banyuls would be worth the journey. Banyuls has an attractive seafront, with statues by Aristide Maillol, who was born here, and Collioure, dominated by its castle, is an enchanting

Roussillon

fishing port with lively streets, known for its anchovies. The appellations stand slightly apart from the rest of Roussillon. The vineyards are even more dramatic than those of the Agly Valley, sitting on steep terraced hillsides, where mechanization is virtually impossible and heroic viticulture is the order of the day.

The city of Perpignan is the focus of the department, and after Barcelona, the second city of Catalonia. I have to admit that I have spent little time there as I have always been eager to reach the vineyards, but it seems to have a different atmosphere from neighbouring French cities. This is a city that is proud to be Catalan. On my last visit, our arrival coincided with a torrential downpour but we managed to take refuge in a friendly pop-up restaurant on the Promenade des Platanes, and then visit the imposing cathedral of St Jean. Nearby are the narrow streets of the old city and to the south, the formidable Palace of the Kings of Majorca, dating from the thirteenth century, when Perpignan was part of that extensive kingdom.

Roussillon does not want for historical sites to visit. There are wonderful abbeys; St Martin-du-Canigou and St Michel-de-Cuxa and the Prieuré de Serrabonne, and more modest, but equally inspiring, the Prieuré de Marcevol. Eus, which rises above the Têt Valley, is classed as one of *les plus beaux villages de France*. The town of Céret has associations with Picasso and other famous artists, as well as a fine museum of modern art, while the nearby chapel of St Martin-de-Fenollar boasts wonderful twelfth-century murals. The cathedral at Elne has evocative cloisters with Romanesque arches and carved capitals. More unexpected is the church in the small village of Baixas which has an enormous 17-metre-high altarpiece covered in gold, demonstrating the financial success of the wine growers of the village when it was commissioned in 1674. Apparently it cost so much that it resulted in a proverb, 'as expensive as the altarpiece of Baixas', to describe something that is exorbitantly expensive. Thuir is a small town that merits a visit, with a welcoming centre dominated by the premises of the once popular aperitif Byrrh. The cellar tour is well worthwhile, and we concluded our visit with what turned out to be our last restaurant meal before 2020's pandemic lockdown, at a friendly hostellerie, Le Patio Catalan, just across the road, that offered a Menu au Byrrh. The *coq au Byrrh* was particularly tasty.

There are many strands to the wines of Roussillon. With some exceptions among the wine estates near the coast close to Perpignan, the vineyards are all on hillsides, some gentle, some much steeper. The flatter land to the west of Perpignan is a vast market garden, above all for apricots and peaches. March is a wonderful time to be there; the orchards are flowering and spring is coming, with vivid splashes of delicate blossom. The climate is essentially Mediterranean, but with climate change it is becoming much less consistent. The winds can blow hard. Winters are usually mild, and summers are hot, with drought conditions increasingly prevalent, having an inevitable impact on yields. As a result of the formation of the Pyrenees, the soil is enormously varied, more diverse than just about any other vineyard of France, with the exception of Alsace. Maury is based on schist, there is granite at Lesquerde and you will also find clay and limestone, marl, sandstone and gneiss. The variations are infinite.

As for grape varieties, Grenache Noir is the key variety for red wine, with the added attraction of old vines. You will see vineyards of gnarled, stubby bush vines that withstand the strong winds. Usually the wines are field blends, featuring all three colours of Grenache, often with some Carignan for good measure. There is growing interest in Carignan, with its acidity providing an important balance to Grenache. Syrah and Mourvèdre also feature, and to a lesser extent Cinsault, which was previously considered too light and not suitable for Vins Doux Naturels. The likes of Cabernet Sauvignon and Merlot are relatively rare in Roussillon.

For white wine, Grenache Blanc and Grenache Gris are important, as is Macabeo, the traditional white grape of Catalonia, which is much less common in the Languedoc. You will also find Carignan Blanc (Carignan Gris is very rare) along with Malvoisie du Roussillon, or Tourbat, which has fallen from favour, but may be in line for a revival. Vermentino, Roussanne, Marsanne and Viognier also feature in the appellations, and you may find occasional examples of Chardonnay and Sauvignon Blanc in the IGPs. Muscat, both Muscat à Petits Grains and Muscat d'Alexandrie, is of course significant for Muscat de Rivesaltes, and also as a vin sec. One of the surprises and discoveries during my research was the quality of the white wines of Roussillon. It is an enigma that so many of the wines simply do not taste as though they come from a hot climate. You could be forgiven for thinking that the climate of Roussillon would be completely unsuitable for white wine, but other factors come into play, such as altitude, the proximity of the mountains and the suitability of the indigenous varieties to the terroir.

The red wines have evolved enormously, with many changes and developments over the years. Winemaking has become more refined, with less heavy-handed extraction, less use of small oak barrels and a shift towards bigger demi-muids and even foudres. You will also find eggs and amphorae. There is a quest for lower alcohol levels. People are experimenting with orange wines and wines without any added sulphur, even wine growers who I would have thought more conventional in their approach than the natural winemakers, for whom any additions or interventions are anathema. The natural wine movement has a firm foothold in Roussillon, centred on the village of Latour-de-France, and is popular with many of the new arrivals in the region. Like the Languedoc, Roussillon attracts outsiders, from elsewhere in France and from other countries and continents. The price of vineyards is such that they are accessible to those with more limited means. Often the newcomers have moved from other fields of activity, bringing a different perspective to a second career in wine. Although production was once dominated by the village cooperatives, these have become very much less important, with an escalation in the number of independent wine estates, each trying to make its mark. Several of the wine estates that I visited have made their first wines within the last five years, and certainly within the past ten years. Work in the vineyards has evolved. Organic viticulture and biodynamic practices are increasingly important, while some growers prefer to follow the requirements of Haute Valeur Environnementale (HVE), which places an emphasis on biodiversity. There is a widespread awareness of climate change and its impact, particularly on yields and grape quality.

Often the appellation requirements are questioned, with the restrictions found to be limiting. Consequently, the principal IGP, Côtes Catalanes, is widely used by those who prefer more flexibility and who find the appellations irksome; the IGP allows for single varietals, which are often particularly successful for white wine. Although blending is the essence of the appellations of Roussillon, sometimes a wine grower does not have the necessary proportions in their vineyards, and will again resort to the IGP (Pays d'Oc is less important in Roussillon than in the Languedoc). And for those who will have no truck with any regulations, there is always Vin de France. Do not dismiss a Vin de France from Roussillon; there will be a good reason why the wine is a Vin de France rather than an appellation or IGP. You may not like it, but it will have been made with passion and commitment.

The most original aspect of wine in Roussillon remains without doubt its Vins Doux Naturels, with the appellations of Maury, Banyuls and Rivesaltes. Rivesaltes and Muscat de Rivesaltes cover virtually the whole department, while Maury and Banyuls are focused on those two villages. The Vins Doux Naturels take two forms: reductive and oxidative. The reductive wines, Rimage and Grenat, are a more recent development, while the oxidative wines are an intrinsic part of the history and traditions of Roussillon. They may be described as *rancio* if they have developed the particular characteristics while ageing, in either a barrel or a glass jar exposed to the elements and extremes of temperature.

However, Rancio Sec, an old tradition that so nearly disappeared, describes a wine that is not fortified, merely aged for several years in barrel without any *ouillage*. Some say that historically *rancio* preceded the Vins Doux Naturels, as *rancio* does not require fortification. It is neither a Vin Doux Naturel, nor a vin sec, but depends on oxidative ageing to develop some wonderful original flavours, not dissimilar to fino sherry. Rancio Sec nearly disappeared, as so few people were making it, but happily a group of fervent enthusiasts managed to stem its decline, and now it features as a category of both Côtes Catalanes and Côte Vermeille.

One of the enigmas of Roussillon is the decline in its Vins Doux Naturels. The best, the Hors d'Age, which have spent at least five years in barrel, are truly wonderful original wines, and yet they have fallen from favour. How can their decline be halted? Another puzzle is why Roussillon has not acquired the cachet of Priorat. My friend and colleague Andrew Jefford describes Roussillon as a northern Catalan echo of Priorat, observing that 'the wines are just as "mineral"; no less overwhelming; often fresher.' I could not agree more.

This book is the result of several recent visits to the region, but my enthusiasm for it began with two much earlier books. My first visit of any length to Roussillon was back in 1987 for *French Country Wines*. Inevitably the village cooperatives featured largely, but I also visited a handful of private wine estates, including Cazes Frères, Château de Jau, Mas Amiel and Château Corneilla as well as other estates that no longer exist. A more extensive visit followed for *The Wines of the South of France* in June 1999, when the highlights included my first visit to Gérard Gauby, who, when I asked him about the history of his estate, replied, with an apology to Louis XIV: 'l'histoire, c'est moi.' I also met Frédérique Vaquer for the first time, and visited other estates that continue to thrive, such as Domaine Cazes, Domaine Piquemal and Domaine des Schistes, and in Banyuls and Collioure, Domaine La Tour Vieille, Domaine de la Rectorie and Domaine Vial-Magnères, all of whom feature in this book.

The research for *The wines of Roussillon* began briefly in the spring of 2018, nearly 20 years later, with an initial visit to Roc des Anges and Domaine Gauby. The object was to introduce some Norwegian friends to the delights of Roussillon, and things got off to a very good start with lunch at Riberach in Bélesta, while we stayed at the Auberge du Cellier in Montner. Research began in earnest in June 2019, with visits in the Agly Valley and a couple of days in Calce. I returned in September, and then in October spent a week in Collioure and Banyuls. My next visit, the following March, was fated, and I returned to London a week earlier than intended as President Macron planned lockdown for France.

My deadline was extended to allow for more visits once travel was possible again, so a full week in July 2020 followed, staying in a cosy gîte at Domaine des Soulanes, and I returned in September for a final couple of days, to tidy up loose ends. Research on the ground could continue almost indefinitely – there is always another lead worth following, another wine grower worth seeing. I know that there are other estates that merit inclusion. Wine growers will happily enthuse with mutual respect about their competitors and you know that a recommendation from one talented wine grower will lead to another. Sometimes a serendipitous bottle led to an enthusiastic cellar visit. Quite by chance on my last evening in Roussillon in late September, we ate at Riberach, with the opportunity to enjoy their extensive local wine list, thus turning full circle.

Altogether this book is the fruit of some 30 days of research on the ground, totalling almost one hundred cellar visits concentrated between June 2019 and September 2020. What follows is the distillation of those conversations and tastings, capturing the current concerns and enthusiasms of the wine growers I talked to. As I was putting the finishing touches to my manuscript, the wines of 2020 were finishing their fermentations and being racked into barrel or vat. Despite the problems and challenges of Covid-19, the wine growers were happy with the harvest.

I would like to give the last word of this introduction to Wendy Wilson of Domaine le Soula, who describes the region as 'a hidden treasure, waiting to be discovered'. So, I would urge you to discover the region for yourselves, first via the pages of this book, preferably with a glass in hand, but I also hope that it will encourage you to visit in person, once we are able to travel freely again.

A note on the text

Some French winemaking terms are much more precise and less clumsy than the equivalent English terms, and the same often goes for wine descriptions. If you are talking to French winemakers, inevitably some French words creep into the notes and the subsequent text. Commonly used terms are explained in the Glossary, which begins on page 265. Throughout, I have anglicized the French word *muté*, meaning stopping the fermentation with the addition of grape spirit, to muted (with apologies to Zoom). There are many spellings of the grape variety Macabeo, including Maccabéo, Macabeu, Maccabeu. I have favoured the Catalan version, Macabeo. Lastly, when I mention Grenache on its own, I mean Grenache Noir, otherwise I indicate which colour of Grenache, Gris or Blanc.

PART 1
BACKGROUND

1

FROM THE GREEKS TO THE TWENTY-FIRST CENTURY

The wines of Roussillon have a long history, as does the region itself, with the oldest remains of European man, some 450,000 years old, found near the village of Tautavel in 1971. The beginnings of grape growing in Roussillon may well date back to the eighth century BC, with vines introduced by Phoenician traders, before the Greeks came here to mine iron ore, leaving evidence of imported wines, in the seventh and sixth centuries BC. It is even claimed that some of Hannibal's warriors settled at Maury, becoming the first wine growers of the region. Hannibal marched from Iberia, over the Pyrenees and then on over the Alps on his way to northern Italy during the second Punic war at the end of the third century BC.

The name Roussillon comes from the Romans and the introduction of winemaking in France is generally attributed to them, rather than the Greeks. Ruscino, the origin of the name Roussillon, was a Roman oppidum, a settlement of some size built on the plain near the banks of the river Têt. It pre-dates Perpignan and its site is now somewhat lost in the city's suburbs. The Romans opened up communications, building the via Domitia, which passed through Rivesaltes, Ruscino and Elne, making the region part of Roman province of Gallia Narbonensis.

In the first century AD the wines of Roussillon were praised by Pliny the Elder, who enthused about the Muscats and mellow sweet wines from grapes gathered on the hillsides and in the gorges of the area, producing a wine comparable to Falernian, Caecuban or the best *crus* celebrated by Horace. The Romans called Muscat à Petits Grains *vitis apiana*, and the grapes were often left to dry on the vine, in order to make sweet wine.

Pliny also commented on the violence of the winds, referring in particular to the circius, which was capable of lifting off roofs. It also forced the wine growers to choose the orientation of their vines very carefully. Cato the Elder, who came to Roussillon in 195 BC, called it the cers. Apparently, although it could blow over an armed man, or full chariot, 'the inhabitants of the Narbonnais were grateful to it, for the healthiness of their climate'. The case is true today amongst the vignerons of Roussillon; although the cers is not one of the winds of Roussillon, there are many others.

With the decline of the Roman Empire came the Visigoths, in AD 719, bringing a period of uncertainty under Theodoric II. The Saracens followed, to be defeated by Pepin the Short in 759, meaning Roussillon was controlled by the Counts of Roussillon, with strong links with their cousins, the Counts of Empúries. In the twelfth century the Counts of Barcelona established their supremacy. In 1172 the last count left his lands to the kings of Aragon, and Roussillon became a buffer between France and the growing power of Castile. With the Treaty of Corbeil in 1258, Louis IX of France formally surrendered sovereignty over Roussillon and his claim to the title of Count of Barcelona to the crown of Aragon. James I of Aragon had seized the Balearic Islands from the Moors and united them with Roussillon to create the Kingdom of Majorca, with its capital in Perpignan. Thus, the region remained under Spanish rule, until it finally became part of France with the Treaty of the Pyrenees in 1659.

In the Middle Ages the church encouraged the development of viticulture. The church at Elne and the first Benedictine monasteries all possessed vines. Roussillon has its share of fine abbeys, such as Serrabone, St Martin-du-Canigou and St Michel-de-Cuxa, and the authority of the magnificent abbey of Lagrasse in the department of the Aude extended into the Agly Valley. The Knights Templar, who first arrived in Roussillon in 1138, brought agricultural innovations, draining lakes and irrigating terraces, planting cereals, olive trees and vineyards. Muscat was an important grape variety; some was sent to the papal court of Benoit XIII at Avignon at the end of the fourteenth century. In the fourteenth century mention is also made of Malvoisie, possibly brought from Greece.

From the thirteenth century onwards Roussillon exported its wines, amongst other goods, particularly through the port of Collioure to the

Balearic Islands and Sardinia, and further east, as well as to Flanders and England. In 1299 James II of Majorca, at the request of his *consuls et prudhommes* at Pepinyà, or Perpignan, forbade the importation by land or by sea of any foreign wine into Roussillon. Pepinyà was accorded a monopoly in the sale of wine in the king's lands. In the fourteenth century Muscat wine was sent by sea and by land to the royal palace in Valencia, and also to Barcelona. There are records that Pierre IV of Aragon in 1345 asked for two tonneaux of the best red wine that could be found at Pepinyà, and his wife, Marie of Navarre, ordered on several occasions the best red wine of Roussillon, 'as Madame la Reine does not wish to drink wine other than which comes from Pepinyà.'

The region's greatest contribution to winemaking was the discovery at the end of the thirteenth century by Arnau de Vilanova that the addition of alcohol to fermenting grape juice brings the fermentation process to a halt. Vilanova, known as Arnaud de Villeneuve in France, a doctor of medicine from the university of Montpellier, is also credited with the introduction of the still from Moorish Spain, producing *eau de vie*, which he called *elixir de vida* and which was also used for medical purposes. The word alembic, or still, is of Arab origin. As will be seen, this discovery was to have a marked effect on the viticultural development of the region. Vilanova, who came either from Valencia or Aragon, was engaged at the beginning of the fourteenth century to prepare a diet for King James II of Aragon. Part of his regime considered drinking, for he believed that man feels two different kinds of thirst: the first is natural thirst resulting from the heat that occurs during digestion, which is alleviated by drinking water; the second is the result of exercise, eating spicy food or breathing dust, and can only be quenched by *vino aguado*, wine blended with water. Vilanova was convinced of the virtue of moderate drinking and his contribution to medicine lasted well into the seventeenth century in Spain, while his contribution to wine is still appreciated.

There are numerous texts from the seventeenth and eighteenth centuries which confirm the reputation of Muscat de Rivesaltes. Muscat was sent to the court in Paris and even as far as Constantinople for Monseigneur l'Ambassadeur de France. The English traveller Arthur Young, writing in 1787, talks of visiting the Lasseuse family at Pia, and seeing their vineyards, where he was told that 'Rivesalta' produced the most famous wine of France, a statement he found to be true at dinner that evening. Voltaire wrote to the governor of the château of Salses of the pleasure that he felt when drinking a glass of the wine of Salses, even though his feeble body was not worthy of this elixir. Grimod de la Reynière, lawyer and gastronome, writing at the end of the eighteenth century, described Rivesaltes-Hautes-Rives as the best *vin de liqueur* in Europe. The wine was served to the King in Versailles, and Reynière noted that Grenache Blanc and Malvoisie added their finesse to the Muscat.

Viticulture has always been the dominant agricultural activity in this part of France. By 1868 Jules Guyot had recorded 60,000 hectares of vines in the department of the Pyrénées-Orientales, observing that the hotter climate, caused by the fact that the land was more sheltered from the wind than in the Hérault, gave wines that were richer, sweeter, more solid and with more colour than those of the Hérault. The vines were planted on terraces and mingled with olive trees. The main grape varieties were Grenache, Carignan, Picpoule Noir, Mataro (the Spanish and Australian synonym for Mourvèdre) Malvoisie, Clairette and Pampanal. Guyot particularly praised Banyuls and the Muscats of Rivesaltes and said that the ordinary wines had a very deep colour, while the fine wines had body, spirit, weight and vinosity. They travelled well and were worth as much as the best wine of Oporto, either as a fortified wine, or mixed with lighter wines.

In his *Ampélographie Française* of 1857 Victor Rendu described the wines of Roussillon, saying that the best-known vineyards were Banyuls-sur-Mer, Collioure, Port-Vendres, Rivesaltes and Perpignan. The Vins Doux Naturels received the most comment, while the table wines or vins secs were credited with little distinction for it seems that their role was in blending. Favoured for their high alcohol and deep colour, they were destined to be mixed with lighter wines that were deficient in those qualities.

According to the Annuaire Statistique of 1866 by Paul Deville, wine was the principal source of wealth for the department. He particularly cites the wine of Rivesaltes as better than the rest, its Muscat and Macabeo enjoying legitimate celebrity status, and also credits Collioure and Banyuls with the greatest reputation of the province. The middle of the nineteenth century saw a significant extension of the vineyards, with Deville giving figures of 38,000 hectares in 1836, rising to 50,000 in 1860 and 57,000 in 1866. Wine was selling well, but then phylloxera, which first appeared in the region in 1877, caused the same devastation as elsewhere in France, and led to the abandoning of hillside sites and development of vineyards on the plain around Perpignan.

Production figures, from Les Vins du Roussillon, show a record 1,752,000 hectolitres in 1881, which dropped to 866,000 in 1885, but rose again in 1888 to 1,121,822 hectolitres. In 1890 there were 30,000 hectares of vines in the department, in other words half the surface area of 1870, and only a third that of 1881. By 1906, replanting or new plantings had brought the area to 65,000 hectares, with a production of 1,797,000 hectolitres. However, 1906 was considered a poor harvest, and proved a catalyst for the viticultural protests that disturbed both Roussillon and the Languedoc, with more than 180,000 people protesting in Perpignan on 19 May 1907. The protests eventually fizzled out after their leader, Marcelin Albert, was outmanoeuvred by the Prime Minister, Georges Clemenceau.

The first cooperatives were created in the region at the beginning of the twentieth century. The first, at Bompas, in 1907, was followed by Estagel, Maury and Espira-de-l'Agly in 1910 and Villelongue de la Salanque in 1911. The movement really took off in the 1920s, with a veritable explosion so that there were more than 60 cooperatives, with one, or even two, for each village, at the height of their importance.

The creation of the aperitif Byrrh[1], which was attributed with medicinal qualities, played an enormous part in the success of viticulture in Roussillon, and also in its decline, when Byrrh fell from favour. In 1866 brothers Pallade and Simon Violet, who were itinerant drapers, settled in Thuir and created the drink, which is based on unfermented grape juice combined with spirit, or *mistelle*, and quinine. The trademark was registered in 1875. In 1910, at the height of its success, the company was producing over 30 million litres a year, and employed 750 people. The focus of production concentrated on Roussillon, with extensive cellars all along the Agly Valley, in Maury, Estagel, Espira-de-l'Agly and other villages. Until the outbreak of the Spanish Civil War they also imported wine from Catalonia. The next generation, Lambert Violet, recognized the power of advertising and one can admire some of his posters at Byrrh's premises in Thuir. Byrrh was described as 'tonique à base de vins généreux et de quinquina'. Bottles were sent to military convalescent homes during the First World War and people were encouraged to drink wine, the national drink, to save France. Hard liquor was forbidden during the war and people turned to vermouth, notably

1 Technically Byrrh should be written BYRRH as the name is made up of the initials of members of the Violet family. However, for ease of reading we have used a more conventional presentation.

Byrrh, so that between the two world wars Byrrh accounted for half the aperitif market of France. Somerset Maugham refers to Byrrh being drunk in Saigon in *The Gentleman in the Parlour*, published in 1930. A tour of the Byrrh premises today includes a huge concrete vat, which holds 680,000 litres and apparently used to take 14 hours to fill, as well as the largest oak vat in the world, a breathtaking 1,000,200 litres, which took 15 years to build.

But gradually sales of Byrrh declined. In 1960, it was absorbed by Dubonnet Cinzano. The last of the Violets, Jacques, died in 1963 and in 1977 the company was acquired by Pernod Ricard. Today its production totals about half a million litres, with two drinks, herb-flavoured Byrrh and Byrrh Grand Quinquina, all drunk mostly in France. The extensive premises, seven hectares in area, are also used for the production of other aperitifs. The vats in the enormous barrel hall are largely empty. It remains a popular tourist attraction in the centre of Thuir, and is worth a visit for its role in the history of the wines of Roussillon. Sadly, the visit does not include the part of the cellars built by Gustave Eiffel in 1892. Old photos show an impressive ceiling of intricate ironwork.

The first appellations of France were created in the late 1930s, an indirect result of the protests earlier in the century, and in Roussillon it was the Vins Doux Naturels that were to benefit from the early decrees. Rivesaltes, Banyuls and Maury continue to survive, but not necessarily flourish despite their stunning quality; Côtes d'Agly and Côtes de Haut-Roussillon have disappeared, following their incorporation into Rivesaltes in 1972. The appellation Grand Roussillon followed in 1956, as a lesser appellation encompassing all the Vins Doux Naturels, but it was of limited success and is now defunct. Muscat de Rivesaltes came relatively late compared to the Muscats of the Hérault, in 1956.

The first appellation for vin sec was Collioure Rouge in 1971; it had originally been called Banyuls Sec. Collioure Rosé and Blanc followed much later, in 1991 and 2003 respectively. In the 1950s, other vins secs were recognized as Vins Délimité de Qualité Supérieure or VDQS, namely Corbières du Roussillon, Corbières Supérieures du Roussillon and Roussillon des Aspres. Then in 1977 the VDQS were rationalized. Corbières du Roussillon and Corbières Supérieures became the appellation Côtes du Roussillon Villages, covering 32 villages in the northern part of the department, with Caramany and Latour-de-France singled out for special mention. Côtes du Roussillon covered the remaining villages, all 118, including Roussillon des Aspres, mostly positioned on flatter land. Further single villages have subsequently been recognized: Lesquerde in 1996 and Tautavel in 1997, and then Les Aspres, first as a terroir of Côtes du Roussillon in 2003, and then as Côtes du Roussillon Villages in 2017, with another 19 villages added to the appellation. There may be others in the pipeline, such as Cases-de-Pène, while Tautavel would like to stand alone, as an *appellation communale* in its own right. Maury has always remained separate, with the appellation for Vin Doux Naturel created in 1936, and for Maury Sec in 2011.

Early in the twentieth century English wine and food writer Morton Shand described the table wines of the Midi in fairly disparaging terms, referring to a classification in the Pyrénées-Orientales of 'Bons Choix, Supérieurs et Petits Roussillons'. He does not elaborate further. When the Algerian war of independence in 1962–3 meant that wine from north Africa was no longer available for blending with the lighter wines of the Languedoc, wine from Roussillon was used instead. Morton Shand also observed that towards the middle of the nineteenth century the wines of Roussillon were especially popular in England, where they were deemed to be the potential rivals of Port on account of their alcoholic strength, 'which it was fondly imagined was *naturally* as high as 27 degrees.' He talks of Banyuls as having a slightly ferruginous taste. Apparently, it was used in most French hospitals and has the distinction of being mentioned in the *Codex*, which corresponds to the *British Pharmacopoeia*. Grenache was deemed particularly suitable for the making of Rancio Sec and the best growths were to be found in Latour-de-France, St Paul-de-Fenouillet, Estagel, Baixas and Tautavel, villages known for the quality of their wines today. Allan Sichel, writing in *The Penguin Book of Wines* in the mid-1960s, talks of Grand Roussillon as an omnibus AOC and singles out four areas, Maury, Côtes d'Agly, Rivesaltes and Banyuls, all Vins Doux Naturels, with Rancio Sec 'roughly equivalent to our Tawny'. He also mentions Corbières du Roussillon as a VDQS.

It is really only in the last 20 years or so that the table wines, or vins secs of Roussillon have acquired a distinctive identity of quality. The change in perception is clearly demonstrated by comparing the entries in the very first and the most recent editions of Hugh Johnson's *World Atlas of Wine*. In the first edition, published in 1971, Languedoc and Roussillon combined had a double page spread. Roussillon is grouped with Corbières, with the observation that 'the vast viticultural zone of Corbières-Roussillon leads a double life, as producer both of medium quality red wines, and of a true speciality, their "vins doux naturels", (which are not as natural as their name makes out, being fortified with alcohol like port).' As Hugh points out, 'the red wines only had VDQS status at that time, but the sweet wines were appellation contrôlée and commanded a vast supermarket sale in France, packed in screw-top bottles,' not, I would add, unlike cheap sherry in England at the time. Fast forward to the eighth edition, published in 2019, with Jancis Robinson MW sharing the work with Hugh Johnson, and we can see what great progress has been made. Roussillon alone commands a double page and there is a detailed appreciation of the quality of both the vins secs and the Vins Doux Naturels, with the pertinent observation that in their heyday in the mid-twentieth century 70 million bottles a year of Vin Doux Naturel were sold, as opposed to barely 3 million today. Such is the sad fall from favour of these wonderful wines.

Perhaps the most eloquent observation I have seen on Roussillon was from Andrew Jefford. Writing in *Decanter* in February 2020 he says, 'Roussillon would be where I'd set about hunting for absolute grandeur …'. I could not agree more, and in the following pages, I hope to explain why.

2

THE APPELLATIONS OF ROUSSILLON IN A NUTSHELL

The Vins Doux Naturels came first with the creation of the appellations of Rivesaltes, Banyuls and Maury in 1936, making them amongst some of the earliest appellations of France. Depending on the exact location of its vineyards, an estate can produce Côtes du Roussillon, Côtes du Roussillon Villages, with or without a named village, and Maury, as well as Rivesaltes and Muscat de Rivesaltes, not to mention the IGP Côtes Catalanes and even Vin de France. It is really only the four villages of the appellations of Banyuls and Collioure, along with the IGP of Côte Vermeille that are strictly defined. With so much overlap, a map of the appellations of Roussillon is quite confusing.

All the Vins Doux Naturels are fortified with neutral grape spirit, in a process called mutage. The decision on when to fortify is not based on a minimum or maximum amount of sugar, but on the balance of the wine, depending on the future category of the wine. There are parameters, but the wine grower has the final say. The volume of alcohol for all the Vins Doux Naturels usually accounts for between 5 and 10 per cent of the total liquid. For Rivesaltes, Banyuls and Maury the lowest permitted sugar level is 45 grams per litre but in practice it is usually 60–70 grams per litre and may be as high as 110–120 grams per litre.

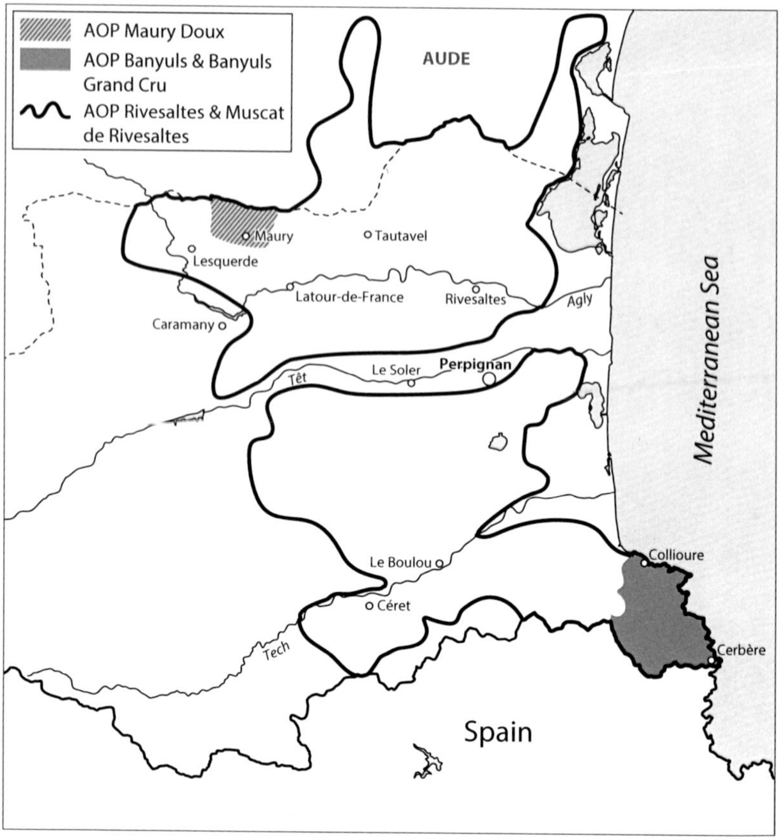

Roussillon vineyards, AOP Vins Doux Naturels

RIVESALTES

Rivesaltes comes from a total of 86 villages in the Pyrénées-Orientales and includes virtually all the wine villages of the region, with a few exceptions such as St Martin-de-Fenouillet, which was deemed to be too cool. It also includes nine villages in the Aude, namely the villages of the inland part of the appellation of Fitou. Rivesaltes is subdivided into four colours; Grenat, Rosé, Tuilé and Ambré, while Muscat de Rivesaltes, created in 1956, stands as an appellation in its own right. Grenache is the backbone of all Rivesaltes. Grenat, recognized as a style of Rivesaltes in 2001, is pure Grenache Noir, whereas for the other wines, Grenache Gris and Blanc also feature, as well as Macabeo, Tourbat or Malvoisie du Roussillon and the two Muscats, à Petits Grains and d'Alexandrie.

The minimum ageing for Grenat, the youngest style of wine, is eight months, and its essence is fresh red fruit. The base wine for Tuilé leans more towards black grapes that turn a brick colour with age, whereas for Ambré white grapes are more important and the wine turns amber-golden with age. Both are aged in an oxidative environment for a minimum of 30 months. With five years or more of ageing, the wine is classified as Hors d'Age. *Rancio* can also apply to Rivesaltes Ambré or Tuilé and describes a wine that develops specific *rancio* characteristics through ageing and oxidation. The lowest permitted sugar level is 45 grams per litre but in practice it is usually 60–70 grams per litre and may be as high as 110–120 grams per litre.

Muscat de Rivesaltes covers 90 villages, the extra four being those of the appellation of Banyuls, covering any Muscat grown in Banyuls. Both Muscat à Petits Grains and Muscat d'Alexandrie are allowed, either as a *monocépage* or blended together in varying proportions. Muscat de Noël describes the first Muscat of the year, which can be marketed from the third Thursday of November. The wine is muted so that it has a minimum sugar level of 100 grams per litre and no more than 130 grams per litre, with a minimum of 15% abv for both Muscat de Rivesaltes and Muscat de Noël.

BANYULS

Banyuls comes from four villages close to the Spanish border: Collioure, Port-Vendres, Banyuls-sur-Mer and Cerbère. It subdivides into five colours: Blanc, Rosé, Rimage (the equivalent of Grenat in Rivesaltes, and a young red wine) as well as Ambré and Tuilé. The principal grape varieties are all three colours of Grenache, with Grenache Noir accounting for at least 50 per cent of the blend, as well as the two Muscats, Macabeo and Tourbat, with Carignan Noir, Cinsault and Syrah recognized as complementary varieties, accounting for no more than 10 per cent. Banyuls *grand cru*, recognized in 1962, must include a minimum of 75 per cent Grenache Noir, and be aged for 30 months in an oxidative environment. In theory, it is a selection of the best wine, but that is open to some discussion (see page 229). For Banyuls the lowest permitted sugar level in the finished wine is 45 grams per litre, and for Banyuls *grand cru* 54 grams per litre, but in practice it is usually 60–70 grams per litre and may be as high as 110–120 grams per litre. The minimum alcohol level is 15% abv.

MAURY

The two appellations of Maury come from four villages, Maury itself, as well as vineyards in Tautavel, St Paul-de-Fenouillet and Rasiguères, with vineyards concentrated in an island of black schist. The colours are Grenat, Blanc, Tuilé and Ambré. For Grenat and Tuilé, the main grape varieties are the three types of Grenache, with Carignan, Syrah and Macabeo as complementary varieties.The Blanc and Ambré are made from the white varieties, Grenache Gris and Blanc, Macabeo, Tourbat and both Muscat d'Alexandrie and Muscat à Petits Grains. For Grenat and Tuilé, Grenache Noir accounts for a minimum of 75 per cent of the blend. Grenat, the youngest and freshest wine, must be aged for a minimum of eight months, while Tuilé and also Ambré are aged for a minimum of 30 months. Hors d'Age entails a minimum of five years of oxidative ageing, and Rancio has the essential characteristics that develop with oxidation. The lowest permitted sugar level is 45 grams per litre but in practice it is usually 60–70 grams per litre and may be as high as 110–120 grams per litre, with a minimum alcohol level of 15% abv.

CÔTES DU ROUSSILLON AND CÔTES DU ROUSSILLON VILLAGES

The appellations for vins secs superseded the VDQS in 1977, with the creation of two appellations, Côtes du Roussillon and Côtes du Roussillon Villages. Côtes du Roussillon comes in all three colours and covers a total of 102 villages. Côtes du Roussillon Villages now includes 51 villages, initially with two named villages, Caramany and Latour-de-France. Lesquerde followed in 1996, Tautavel in 1997, and Les Aspres, which was first attached to Côtes du Roussillon in 2003, became Côtes du Roussillon Villages in 2017. Blending is the name of the game, with a diversity of different grape varieties, in differing percentages. The permitted grape varieties for the red wine are Grenache Noir, Carignan Noir, Lledoner Pelut, Cinsault, Syrah, Mourvèdre and Macabeo, with Grenache Gris allowed for rosé. For red and rosé, there must be a minimum of two grape varieties, with no more than 70 per cent of the dominant variety. The maximum amount of Carignan is 50 per cent and Syrah and Mourvèdre must account for no more than 25 per cent, either together or separately. For rosés too, there must be a minimum of two varieties, with no more than 70 per cent of one variety. The permitted yield is 48 hectolitres per hectare.

For white wines, the permitted grape varieties are Grenache Blanc and Gris, Macabeo, Tourbat, Roussanne, Marsanne, Vermentino, Viognier and Carignan Blanc. There must be a minimum of two varieties, with the main variety making up no more than 80 per cent of the blend. Grenache Blanc, Macabeo and Tourbat, together or separately, must account for at least 50 per cent of the blend, while Viognier and Carignan Blanc must not exceed 10 per cent.

Côtes du Roussillon Villages covers 51 villages in the northern part of the department, limited by the Aude, the étang de Salses, the river Têt in the south and the Fenouillèdes hills to the west. The vineyards lie at altitudes of between 100 and 400 metres. The wine is only red and the grape varieties are Carignan Noir, Grenache Noir, Lledoner Pelut, Syrah and Mourvèdre. There must be a minimum of two varieties, with the main grape variety making up at least 70 per cent of the blend. Syrah and Mourvèdre must not account for more than 30 per cent of the blend, either together or separately. The permitted yield is 45 hectolitres per hectare. There are subtle differences between the various named villages.

Caramany covers three villages; Caramany itself and neighbouring Cassagnes and Bélesta, with vineyards at an average altitude of 250 metres. Wines must contain a minimum of two varieties, with the principal grape making up between 40 and 70 per cent of the blend. This is a village where carbonic maceration was once important, thanks to a technically advanced cooperative in the 1980s.

Lesquerde covers that village, as well as Lansac and part of Rasiguères. Permitted varieties are Carignan Noir, Grenache Noir, Lledoner Pelut and Syrah, but not Mourvèdre. Again, there must be at least two grape varieties in the blend. Carbonic maceration should be used for Carignan and it must not account for more than 60 per cent of a wine.

Latour-de-France covers Cassagnes, Montner, Estagel and Planèzes as well as Latour-de-France itself, with the customary Grenache Noir, Carignan Noir, Lledoner Pelut, Syrah and Mourvèdre permitted. A minimum of two varieties must be used, and Syrah and Mourvèdre must account for at least 30 per cent of the blend, either together or separately. The main variety should not be more 70 per cent of the blend.

Tautavel covers the village of Vingrau as well as Tautavel itself, with the usual five varieties allowed. The usual limit for the main variety is 70 per cent and there must be at least 20 per cent Grenache Noir and/

or Lledoner Pelut. Syrah and Mourvèdre must account for 30 per cent of the blend and there should be no more than 60 per cent Carignan.

Les Aspres covers 19 villages in the heart of the department, south-west of Perpignan, with a carefully delimited vineyard area and even more complicated blending requirements. There must be a minimum of three varieties, with the two most important accounting for no more than 90 per cent of the blend. Syrah and/or Mourvèdre must account for at least 25 per cent. Syrah, Mourvèdre and Grenache Noir must individually not account for more 50 per cent and Carignan must not exceed 25 per cent. The yield for all the named villages is 42 hectolitres per hectare.

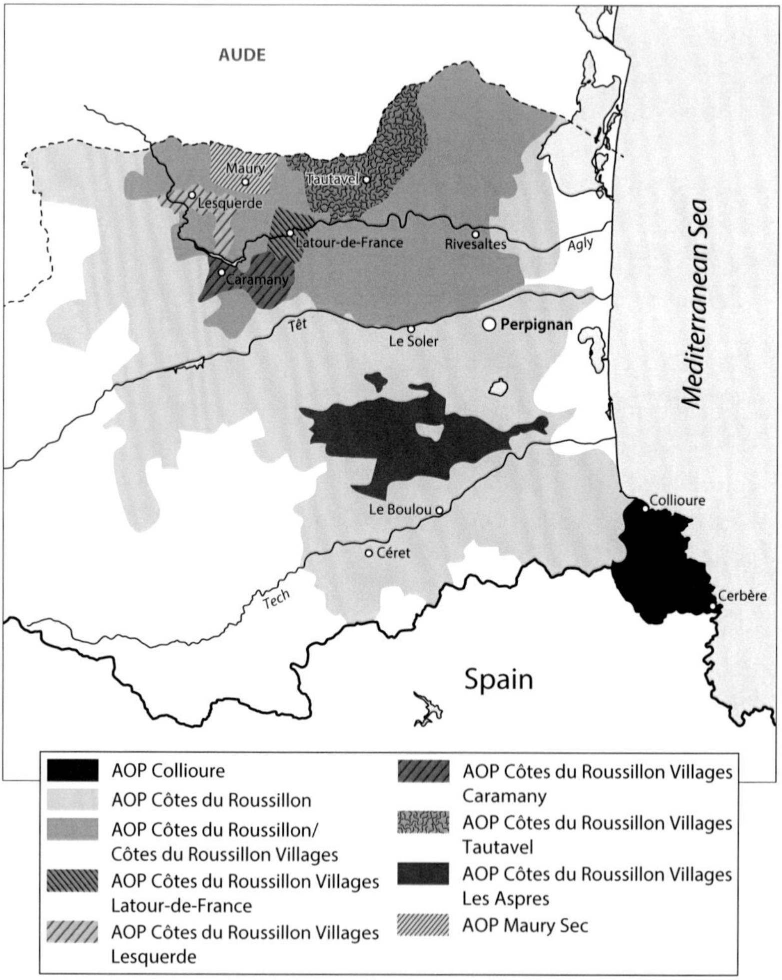

Roussillon vineyards, AOP dry wines

MAURY SEC

The appellation of Maury Sec was recognized in 2011. The vineyards are identical to those for the Vins Doux Naturels. The wine grower decides which wine they will make, depending on the quality of the grapes. Grenache Noir is the main variety, with Carignan Noir, Mourvèdre and Syrah considered complementary varieties and Lledoner Pelut an accessory variety. There must be a minimum of two varieties, and Grenache Noir must account for at least 60 per cent of the blend, and 80 per cent of the vineyard area. Lledoner Pelut must be no more than 10 per cent of the blend. The wine must be aged for at least six months, but in practice the élevage is often longer. The permitted yield is 40 hectolitres per hectare.

COLLIOURE

Collioure covers the same area as Banyuls and again the wine grower decides which wine to make, depending on the conditions of the year. The red appellation was created in 1971, rosé followed in 1991 and white was only added in 2003. The grapes for red and rosé are Grenache Noir, Syrah and Mourvèdre, with Carignan Noir, Cinsault, Grenache Gris and Counoise considered complementary varieties. For white wine, the main varieties are Grenache Blanc and Gris, with Macabeo, Malvoisie, Marsanne, Roussanne and Vermentino considered complementary varieties. The principal varieties should account for a minimum of 60 per cent of a red blend, and Grenache Blanc and Gris should account for at least 70 per cent of a white blend.

OTHER DESIGNATIONS

Given the complicated and quite restrictive blending requirements, it is hardly surprising that many wine growers prefer to opt for an IGP or even Vin de France. Originally there were several different vins de pays: Coteaux des Fenouillèdes, Les Haut Fenouillèdes, Côte Catalane, Val d'Agly and the department of the Pyrénées-Orientales. In 2011 they were rationalized and replaced with IGP Côtes Catalanes, which covers 118 villages in the department, in other words all the wine villages except those of the Côte Vermeille. The permitted grape varieties for red and rosé are Grenache Noir, Carignan Noir, Cinsault,

Lledoner Pelut, Mourvèdre, Syrah, Merlot, Cabernet Sauvignon, Cabernet Franc, Chenanson and Marselan. I came across a couple of instances of Marselan, and just one of Chenanson, at Mas Llossanes. For whites the choice is Grenache Blanc and Grenache Gris, Macabeo, Tourbat, Marsanne, Roussanne, Vermentino, Muscat d'Alexandrie, Muscat à Petits Grains, Chardonnay, Sauvignon Blanc and Viognier. Note the intrusion of more international varieties, which are kept out of the appellations. Since 2012 Côtes Catalanes has also allowed Rancio Sec. The wine requires a minimum of five years oxidative ageing without any topping up, and can only be made from Grenache Noir, Grenache Blanc, Carignan Noir, Carignan Blanc, Cinsault, Mourvèdre, Macabeo, Tourbat, Muscat d'Alexandrie and Muscat à Petits Grains.

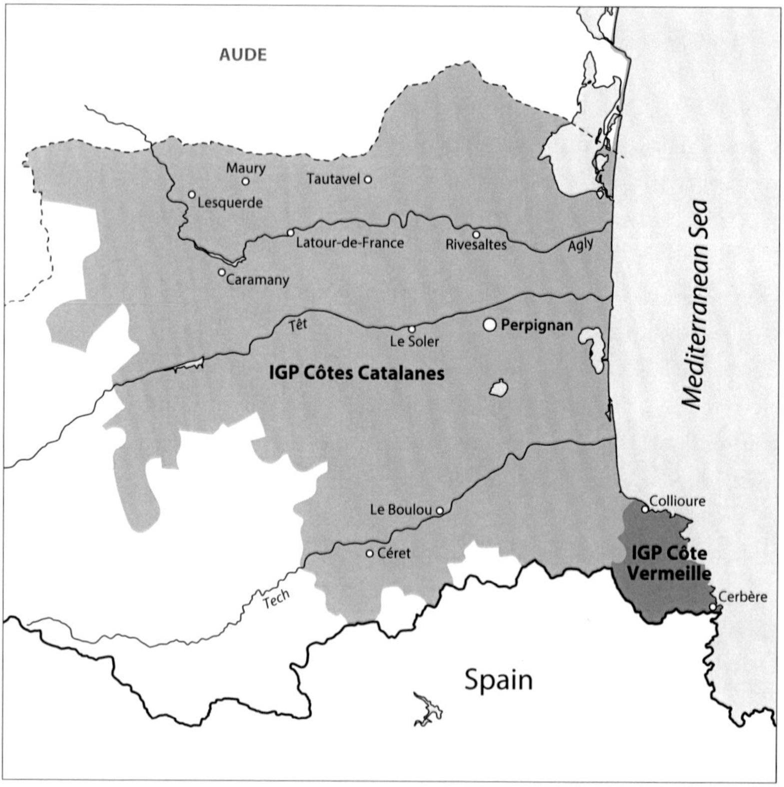

The IGPs of Roussillon (Pays d'Oc can be produced across the department)

Since 2020, Côtes Catalanes has also allowed the use of five disease-resistant grape varieties, namely white Muscaris, Soreli and Cabernet

Blanc, along with Souvignier Gris, and Cabernet Cortis, which is a red variety. The maximum percentage for these varieties is 15 per cent, but for the moment their use is very much experimental.

Côte Vermeille covers the four villages of Banyuls. The permitted grape varieties are the same as for Collioure. The maximum yield is 80 hectolitres per hectare as opposed to 90 hectolitres per hectare for Côtes Catalanes and 40 hectolitres per hectare for Maury and Collioure. In reality, these yields are never reached in the dry conditions of Roussillon. Rancio Sec is also part of the IGP.

The final IGP of Roussillon is Pays d'Oc, as Pyrénées-Orientales is one of the four departments included in that extensive Mediterranean IGP. It allows for a very broad choice of grape varieties, 58 at the last count, as well as the five disease-resistant varieties mentioned above. However, most wine growers prefer Côtes Catalanes to Pays d'Oc, as it is so much more specific to the region.

3

HEROIC VITICULTURE

There is no doubt that Roussillon provides some extremely challenging terrain for vineyards, especially in Collioure and Banyuls. Not for nothing was the expression 'heroic viticulture' coined. Eighty per cent of the vineyards of the region are on hillsides. However, there is nothing particularly unusual about viticultural practices here as compared to any other wine region of the south. The grape varieties are similar, but with a difference of emphasis. Grenache in all three colours features largely, and Cinsault rarely. There is Macabeo, which you hardly ever find in the Languedoc, while Tourbat, or Malvoisie du Roussillon, is only found in Roussillon. Maybe here there are more vineyards of centenarian vines, wonderful old gnarled bush vines. The wind is a significant consideration; it can blow hard and break feeble vine shoots, but it also protects against disease, dispersing humidity after storms. Like wine growers everywhere these days, those in Roussillon are very aware of the effects of climate change. The pattern of rainfall is no longer regular, and in recent years, there has been much less rain. Accordingly, harvest dates are earlier, and yields, never large, are shrinking. Organic and biodynamic viticulture are growing in importance, as they are elsewhere, and it may be that Roussillon has more than its fair share of natural winemakers.

Essentially the climate of Roussillon is Mediterranean, so winters are mild and wet, while summers are warm and dry. It has some of the longest hours of sunshine of any vineyard of France, registering a yearly average of 2,530 hours of sunshine, with an average of 500–600 millimetres of rain. As will be seen, the wind is a key aspect of the climate of the region. At times it can be beneficial, and other times not.

The whole region was formed by the creation of the Pyrenees, throwing up an incredible mix of soils, so that the vineyards of Roussillon are some of the most varied in the whole of France. The one soil type that you do not find in Roussillon is volcanic soil, but limestone, clay, gneiss, various schists and sandstone are all there somewhere. The soil variations are infinite. Granite reflects heat while schist retains it. The limestone of Vingrau gives density to the wines, while Tautavel has more acidity in the soil than Maury. Above all, among growers there is a search for cooler sites, with the prevailing quest for freshness. More differences are discussed in the context of the various wine growers' own vineyards.

GRAPE VARIETIES

Eric Monné of Clot de l'Oum is convinced that the future reputation of Roussillon will be based on the quality of its white wine, with the principal grape varieties of Grenache Gris, Carignan Gris and Macabeo assuming the greatest importance. Grenache, or Garnacha as it is called in northern Spain, loves the hot, dry conditions of Roussillon. With its potentially high alcohol levels, it is the ideal variety for Vins Doux Naturels. **Grenache Gris**, with its pale pink skin, is generally considered to produce wines of greater quality than **Grenache Blanc** and will feature more often as a varietal wine. As well as aroma, and breadth of expression, Grenache Gris has acidity, while Grenache Blanc can be quite alcoholic, making it more suitable for Vin Doux Naturel. In the vineyard the two varieties behave in much the same way as Grenache Noir.

Macabeo originated in the Penedès region of Catalonia, travelling north across the Pyrenees at the end of the eighteenth century. It is now well established in Roussillon, and is indeed becoming quite fashionable, with its neutral vivacity. It buds later and ripens late, dislikes the damp, and responds well to low yields and early picking, which retains its acidity. Marjorie Gallet at Roc des Anges considers Macabeo a wonderful grape: 'People say it is too neutral but it expresses soil differences. It may be difficult to harvest; you must get the right date – the difference between too early and too late can be just two days. You must check it every day; its vibrancy is linked to the low pH.' She makes two different pure Macabeo wines, Pi Vell and L'Oca, while most people blend it with other varieties. You will also find convincing examples of pure Macabeo at Mas Karolina and Domaine La Toupie. Daniel Laffite at Domaine des Soulanes also like Macabeo as 'it respects the terroir'.

It can enhance a blend of Grenache and Carignan Blanc, and also responds well to élevage in barrel.

Frédérique Vaquer if Domaine Vaquer is replanting **Carignan Blanc**, which adds freshness to Grenache Gris and Grenache Blanc. Carignan Blanc, like Carignan Noir, buds late and ripens late, and retains its acidity. You can find old vines of Carignan Blanc, and also Carignan Gris, mixed up with Carignan Noir. Planted in the middle of a vineyard of Carignan Noir and vinified as a field blend it can add minerality and acidity to the wine. Usually it is blended with Grenache and Macabeo but there are occasional examples of pure Carignan Blanc from, for example, Mas Lavail and Domaine Modat. Certainly it is a grape variety that deserves more attention. The same goes for **Carignan Gris**, which is a mutation of Carignan Noir, but there is very little planted. The Danjou brothers are unusual in producing a pure Carignan Gris, as is Domaine Riberach.

Vermentino, or Rolle, is a relatively recent arrival, first appearing at the beginning of the 2000s, and it is working well, especially as it resists drought, and is liked for its acidity. Bud break is early but it ripens slowly and retains its freshness. It can be difficult to control the ripening as it is not always very even, with golden grapes hiding green ones behind them in the bunch, so you can have potential alcohol ranging between 11 and 14% in the same plant, depending on the exposure to sunlight, making maturity difficult to control. Usually it is a small part of a blend with Roussillon's more traditional varieties. However, Robert Pouderoux of Domaine Pouderoux in Maury makes an appealing pure Vermentino.

Roussanne is another recent arrival that, like Marsanne, is included in the white wine appellations. Alain Razungles of Domaine des Chênes talked about discovering it with a member of the Jaboulet family from Hermitage. Frédérique Vaquer is another supporter of the variety. However, Pierre Gaillard of Domaine Madeloc found it disappointing in comparison with its performance in the northern Rhône. It has been criticized for being too honeyed, and the same goes for Marsanne. Both tend to perform best in a blend.

Victor Gardiés of Domaine Gardiés discussed **Tourbat**, or Malvoisie du Roussillon, which is the same as Torbato of Sardinia. According to Victor, everybody had pulled it up in favour of Macabeo, as it is fragile and also susceptible to *court noué*. However, it is a late ripener, even later than Carignan Blanc, and retains freshness, giving tension to a wine, as it does in Victor's La Torreta, where it is blended with 15 per cent Macabeo to fill out the middle palate. Jean-François Deu at Domaine du Traginer is another Tourbat enthusiast, having taken cuttings of it from Etienne Montès at Domaine Cazenove some 15 years ago. With its firm natural acidity, it is now beginning to attract attention again.

Gérard Gauby of Domaine Gauby has planted **Bourboulenc**, which he calls Blanquette and which is more commonly found in La Clape, observing that 'les vieux' hung it in the attic to provide raisins for Christmas. As the emblematic variety of La Clape, there is no reason why it should not also perform well in Roussillon.

Muscat à Petits Grains and **Muscat d'Alexandrie** are of course both extensively grown for Muscat de Rivesaltes, and nowadays also for dry Muscat, but the vineyard area devoted to them (see Appendix III) has decreased dramatically in recent years, owing to the decline in popularity of Muscat de Rivesaltes. Muscat à Petits Grains is generally considered to be the better quality of the two. A very old grape variety, grown extensively around the Mediterranean, it probably originated in Greece and was brought to the south of France by the Romans. It buds early, and can withstand drought. As the name implies, its berries are smaller than those of Muscat d'Alexandrie, which has large bunches of big berries, with higher sugar levels. Muscat d'Alexandrie has nothing to do with the city of the same name, and recent DNA analysis suggests that its parentage is a cross between Muscat à Petits Grains and an obscure red table grape from Sardinia, Axina de Tres Bias. Most often, but not always, the two varieties are blended for Muscat de Rivesaltes, with Muscat à Petits Grains giving some lemony flavours with more elegance, while Muscat d'Alexandrie is riper with more body. However, Muscat d'Alexandrie ripens three to four weeks later than Muscat à Petits Grains, which can make blending problematic. Both cope admirably with the drought conditions that are becoming a more frequent feature of Roussillon viticulture.

Chardonnay is of little significance in Roussillon. The cooperative in Rivesaltes claims to have been the first to plant Chardonnay in the region, in the early 1980s, but today really prefers to concentrate on traditional varieties. Gérard Gauby has a little Chardonnay that is blended into Les Calcinaires. Luc Devot at Débit d'Ivresse makes an intriguing Macabeo–Chardonnay blend and you will find other pockets of Chardonnay, often on flatter land, for IGPs, or at the cooler end of the Agly Valley.

Wendy Wilson at Domaine le Soula and Marc Bariot at Clot de l'Origine were the only people I met who were taking **Sauvignon Blanc** seriously, but their vineyards are cooler than most. Simon Dauré at Château de Jau, who has Chardonnay, considers that Sauvignon is not worth the bother.

As for red wine, the emblematic variety is indisputably Grenache Noir, which has long been the mainstay of the vineyards of Roussillon. The old vineyards are usually *complanté* with Grenache Gris and Blanc, and also Carignan. The closely related Lledoner Pelut is allowed for vin sec, as a *cépage accessoire*, but not in Vin Doux Naturel, and only in small quantities. Syrah and Mourvèdre are the most recent arrivals; there is Counoise too, but very little Cinsault. The likes of Cabernet Sauvignon and Merlot do not really contribute to the flavours of Roussillon; you will find them in IGPs.

The origins of **Grenache Noir**, or Garnacha, are probably Spanish, from Aragon – though Sardinia is suggested as another possible source. According to Robinson, Harding and Vouillamoz in *Wine Grapes*, their exhaustive survey of grape varieties, it arrived in Roussillon in the late eighteenth century and spread throughout the south of France. Enthusiasm for Grenache abounds. Pierre Gaillard considers Grenache to be the Pinot Noir of the south. Charles Perez talked about growing it on sand, where it ripens without too much alcohol. Clay makes for higher alcohol as the soil is warmer and you obtain an alcoholic ripeness, but not phenolic ripeness, but with sandy soil, Grenache is ripe at 14% potential alcohol, with ripe pips. To compensate for the heat, Charles is planting on north-facing sites. At 15% potential alcohol, Grenache is more aromatic, but obviously has less acidity. The principal defect of Grenache, that it is too alcoholic, also makes it suitable for Vin Doux Naturel. You must not pick it too early; the grapes must be ripe and have body to balance the alcohol.

One significant problem for Grenache in the vineyard is *coulure*, if the weather is problematic, with wind or rain, at flowering. Jacques Montagné at Clos del Rey explained how they delay pruning the Grenache so that it flowers when the weather is more stable, and hopefully suffers less from *coulure*. They tidy up the vines in December, and then prune properly leaving two 'eyes', or buds, at the end of March.

Lledoner Pelut is a variant of Grenache, with smaller berries, thicker skins and more acidity, while 'pelut' describes the furry down on the underside of the leaves, which Grenache does not have. It may also be more vigorous and less prone to *coulure*, and copes well with strong winds in the exposed vineyards of Roussillon. It reaches lower alcohol levels than Grenache and is therefore less suitable for Vin Doux Naturel. Usually it is *complanté* and blended with Grenache, as well as Carignan. However, amongst others, Tom Lubbe at Domaine Matassa makes a pure Lledoner Pelut, Cuvée Romanissa, describing it as the Catalan Grenache, and Ronald Joachin's pure Lledoner Pelut is appropriately called C'est un cépage catalan.

Carignan, the Mazuelo or Cariñena of north-eastern Spain, with smaller berries, thicker skins and more acidity than Grenache, can be the best and the worst performing grape here. For Bruno Ribière of Domaine Ferrer-Ribière, Carignan is the *cépage roi* for vin sec, since it gives freshness and acidity, indeed as much acidity as white grapes, and provides backbone in a blend, balancing the riper fruit and alcohol of the Grenache. It is more productive than Grenache Noir and can be planted in richer soil at the bottom of a slope. It is sensitive to oidium. However, old vine Carignan, with lower yields, is more interesting. For Jean-Marc Lafage of Domaine Lafage, Carignan is 'a little treasure'. Old vineyards of pure Carignan do not exist; in the past, it has always been *complanté* with all three colours of Grenache, and in Banyuls probably accounts for about 10 per cent of any old vineyard. It stops Grenache from oxidizing, giving a wine structure and a fresh finish. Some people, such as Olivier Pithon of Domaine Olivier Pithon, are beginning to replant it, especially on poorer soil. Olivier likes it for its elegance and finesse, although it can be astringent.

Thomas Teibert of Domaine de l'Horizon criticized the INAO for insisting on a percentage of **Syrah** in the appellations, as a *cépage améliorateur*, as though Grenache and Carignan are worthless. Bruno Ribière talked of the 'intrusion' of Syrah in Roussillon. He considers that it would be very much better to concentrate on Carignan and to decide which terroir suits that best, and the same goes for Grenache. Syrah is not *à sa place*, he feels. For Pierre Gaillard, there is no Syrah as good as that of the northern Rhône, or even Faugères, where he has another estate, Domaine Cottebrune. A lot of Syrah was planted in the 1980s, by cooperative members, but not necessarily the best clones, nor in the best sites. Syrah often needs irrigation to alleviate water stress. Jean-François Rière of Domaine Riére Cadène uses drip irrigation and really sees the difference compared to vines that have not been irrigated. He observed that some people are picking Syrah at 19–20% potential alcohol as they want ripe phenolics, whereas he prefers to pick at 14.5–15%, just before the phenolic ripeness. Syrah fares particularly badly in hot, dry conditions as it loses its juice and overripens. There are also problems with high mortality rates in the vineyards, with Paul Meunier of Domaine Paul Meunier Centernach commenting that there is very little Syrah in

Maury, thanks to drought and vine trunk disease. Essentially they believe Syrah should not be imposed on the appellation regulations.

Mourvèdre undoubtedly fares better. In Collioure particularly it benefits from the proximity to the sea. Pierre Gaillard finds it capricious, but when it is good, it is very good, giving spice to a wine. It is a late-ripening variety and is less alcoholic than either Grenache or Syrah. Robert Pouderoux described it as having a calming influence on Grenache, acting as a counterbalance, and thought it was being rediscovered. It is quite different from the Mataro of Catalonia. Jean-Roger Calvet of Domaine Thunevin-Calvet would like to try a pure Mourvèdre, as he appreciates it for its elegance. Gérard Gauby considers Mourvèdre to be better than Grenache, while Syrah is too easy. Gérard is also enthusiastic about Roubia, an old variety that was new to me, which is apparently close in character to Mourvèdre. He has a vineyard that he has grafted with massal selection.

Traditionally there has been very little **Cinsault** in Roussillon. It is not suitable for Vin Doux Naturel as the yields are too low and it performs poorly on schist. For Robert Pouderoux it is simply not a variety of Roussillon. However, Olivier Pithon is considering planting some, for its elegance and finesse. Jean-Philippe Padié of Domaine Padié, his neighbour in Calce, has planted some, as have Gérard Gauby, experimenting with it on its own roots, and Mickaël Sire of Domaine des Schistes, but with the intention of making some frivolous rosé. Mas Amiel is also experimenting with Cinsault and I sense a growing interest in a grape variety that copes well with drought conditions.

Counoise features in the *cahier des charges* of Collioure, but not in Côtes du Roussillon or Côtes du Roussillon Villages. In the late 1970s Dr André Parcé planted some cuttings of Counoise he had recovered from the Rhône Valley, and Domaine de la Rectorie and Mas Blanc each have a 70 ares plot. It is part of the blend of Domaine de la Rectorie's Collioure Montagne, but to my knowledge no one else has planted it.

SCIENTIFIC DEVELOPMENTS FOR GRAPE VARIETIES

I spent an interesting and illuminating couple of hours talking to Julien Thiery who manages the research station at Tresserre. The research station is in the middle of nowhere, with no neighbours, so here they are able to plant vineyards in virgin soil, knowing that the soil is virus free. They have what Julien called three pillars to their research, of which two are relevant to this chapter, namely grape varieties and vineyard management. Initially they focused on Vin Doux Naturel, so that they worked on all three colours of Grenache, as well as Muscat and Macabeo, looking for clones that produce lots of sugar. These days, with the growth in vin sec, the opposite is the case. Research is focused on Syrah, and why it has such a short life span. Is there a genetic factor, or does it relate to rootstock or clonal selection? They have found new clones which will counteract its decline, but for the moment these are insufficiently productive.

Research into what the French call *cépages tolerants*, varieties that are resistant to disease, with the aim of reducing the use of chemical products in the vineyards, is another significant area in which considerable progress is being made. New varieties that are resistant particularly to mildew, and to a lesser degree oidium, are coming into production for IGP wines, namely white Solaris, Muscaris and Cabernet Blanc, Souvignier Gris and red Cabernet Courtis. So far there are about 60 hectares of Souvignier and Muscaris planted in Roussillon.

The researchers are also looking at grape varieties that might be less affected by climate change, namely varieties from warmer countries. Julien gave me a long list, as follows: for reds, Agiorgitiko, Aglianico, Zinfandel (or Primitivo) and Bobal, and for whites, Assyrtiko, Airen, Trepat, Malvoisie de Sardaigne and Torrontés. In 2014 they planted 100 vines of each variety, dividing each between three different soils, to see how they compare to Grenache Noir or Grenache Blanc. Above all they are looking at their reaction to drought conditions and are doing micro-vinifications to assess quality. So far it is much too early to make any clear assessment, but Julien did admit that Assyrtiko was looking promising. Alcohol levels are included in their research, especially for Grenache Noir and Grenache Gris.

In the vineyard, the team is looking to reduce the number of chemical products used while at the same time controlling disease. Vineyard management therefore entails organic viticulture, working the soil to remove weeds and suppressing the use of herbicides. *Enherbement* or ground cover is a consideration, but Julien commented that with grass, a vineyard is no longer a natural firebreak, a potential downside. Biodiversity is another factor under consideration; creating habitats for bats, installing bee hives, and cultivating aromatic plants all help to maintain a healthy ecosystem.

AGRICULTURE FORESTIERE

Julien does not have enough space to conduct research into *agriculture forestière*, or forestry agriculture, the planting of trees and hedges in and around vineyards to help growing conditions. Consequently, for insights into that subject, I am indebted to Gérard Gauby, for that is what really interests him these days. Since 2000 he has planted 5,000 trees: cypresses, almonds, olives, oaks, peaches and apricots. Altogether he has about 700–800 olive trees punctuating his rows of vines, with hedges every 16–18 metres. Trees from the Fabaceae family, which includes willow and argania, for argan oil, have also been planted. These have a deep root pivot, which seeks water, and absorbs nitrogen, so when they lose their leaves, the soil is enriched. Gérard is adamant that contrary to what many people think, it is not trees that cause vines to die. The old vineyards were always surrounded by trees; instead of competition he talks about symbiosis.

Gérard is deeply concerned about global warming, and also worried about how to limit the impact of the wind, which blows so hard in Roussillon. There is not just one wind, but as many as eight. The tramontana, or tramontane comes from the north-west; the canigonenc from the west. The wind from the south-west, coming from Spain, is called the vent d'Espagne, while the wind from the south is the sirocco, and the south-east wind, coming off the sea, is the marinada. The east wind is the llevan, the north wind the mistral, and the wind from the north-east, from the direction of Narbonne is called the narbonnais, grégal or gargal. The marinada brings moisture, whereas the tramontane is always dry and can be very hot. Speeds of 100 kilometres per hour are relatively frequent, and 200 kilometres per hour is not unknown. A record 214 kilometres per hour was registered at Cerbère on 26 May 2003.

Wind evaporates moisture, whereas hedges help to retain humidity. With double hedges, 16 metres apart, the wind does not touch the soil. Hedges, if they are high enough, will also create shade, and protect the vines during the winter, making for milder conditions. Gérard observed that with *agriculture forestière*, the ambient temperature of the vines is 4–6°C cooler in summer and winters are milder.

COVER CROPS

Gérard is passionate about *enherbement* or cover planting in the vineyards, and explained that he uses plants from the Fabaceae family, such as lucerne (alfalfa), which enrich the soil with nitrogen and help guard against erosion. The plants are sown in August, and mowed the following May. A legume from the same family is the *trèfle incarnat* or crimson clover, which is particularly suitable as it likes the acid soil of Roussillon, whereas lucerne does better on limestone. Its roots can go as deep as four metres. Each plant has its role. Other cover crops include various grasses, cereals and old varieties of wheat, such as rye and spelt, and crucifers, such as mustard, rocket (which often grows naturally) and different varieties of radish, as well as wild flowers.

Gérard inherited a vineyard from a cousin, and planted lucerne to decompact the soil, so that there was no more *ravinement* or gully erosion, with the water filtering naturally through the soil. He believes in leaving land fallow for as long as eight years before replanting. It is a grave error to plant too quickly, he says, as it means you have to disinfect the soil. A north-east facing vineyard comprised six rows of vines, and then a row of trees. It was already surrounded by trees and *garrigue*. During the winter months about 200 sheep roam his vineyards. They spend the summer months in pastures in Haute Provence – apparently, there is no suitable pasture for them in the Pyrenees. Wendy Wilson at Le Soula also has sheep in the vineyards during the winter, as does Domaine Riberach.

Charles Perez at Mas Bécha makes his own compost, using fresh manure, grape skins and fruit from his neighbouring orchards. He too is experimenting with ground cover, trying out various different mixes; a farmer who grows wheat on part of Charles's land, provides the grain. It is all a question of restoring organic matter to the soil, he says.

Tom Lubbe, who is among the pioneers of organic viticulture in the region, has numerous thoughtful observations on the subject. He does not plough much, preferring what is called 'ripping', a regenerative soil farming practice that digs in precisely the area where vines are to be planted, going 40–100 centimetres deep, to break up soil compaction. He sows cover crops immediately after the harvest, which makes for more nitrogen than sowing a couple of months later. Bare soil is not a good idea, as it encourages evaporation. With cover crops you can lose as much as 2–2.5% abv in the wine, which helps address one of the current preoccupations, how to limit alcohol levels. Tom refuses to accept that 'the sun is pushing us to bigger wines'.

VIEILLES VIGNES AND COMPLANTATION

Old vines with *complantation*, or field blends, is an intrinsic part of the history of Roussillon. You will often find all three colours of Grenache in the same vineyard, as well as some Carignan. Gérard Gauby must have some of the oldest vineyards in the region, with vines planted before 1870, but there are others. All his grandfather's vines were *complanté*, but after the war field blends disappeared. However, Gérard favours *complantation* and thinks nothing of having all three colours of Grenache in the same vineyard. His oldest vines from his grandfather are 130 years old; the 1890 *cadastre* shows Carignan, Grenache, Macabeo and Tourbat. He also has some vines that were planted before 1870, pre-phylloxera Grenache and Macabeo, which the previous owner sold to Gérard, 'as he knew I would not pull them up.'

At Domaine de Riberach they have 80- to 110-year-old vines. They use massal selection for replanting, grafting in the vineyards. The rootstock is planted first and grafting takes place three years later. It can be very difficult to replace single vines, with the problems of virus and disease. For Cédric Haague, a winemaker at Vignerons Catalans, there is no doubt that the old vines are the trump card of Roussillon, even though the vineyards are not necessarily very profitable. However, most agree that they enhance the quality, with more concentration and depth of flavour, a result of their deep root systems, which also enable them to cope with drought conditions.

While most extol the virtues of old vines, a more cynical observer noted that while you are allowed 6,000 kilos of grapes per hectare, you can get as much as 15,000 kilos from one plot, but achieve the correct crop levels per hectare by including the old vines in the next-door plot in your total hectarage, even though they may not have produced a crop at all.

MECHANIZATION

Many of the vineyards of Roussillon are on steep slopes where mechanization can be difficult, and on the terraces of Collioure and Banyuls it is well nigh impossible. They might be wide enough to allow for narrow caterpillar tractors, but nothing more powerful. Consequently, most of the work is done by hand. Weedkiller was used in the 1960s, but these days every responsible wine grower tills the soil. Grass helps guard against erosion, which can be a problem when it rains heavily. It will be cut, or may provide grazing for sheep. Terracing itself provides some damage limitation against erosion. Bruno Duchêne at Les 9 Caves used a horse for ten years, but now prefers a small caterpillar tractor. Victor Gardiés uses a horse in some of his small plots, which are very difficult to cultivate. Paul Meunier observed that in his oldest vines, with very narrow rows, a tractor is impossible; again, he can only use a horse. Ploughing can create dust and lead to water stress, and an intercep is often used for weeding between the vines within a row. Inevitably old vines require more manual work. Cyril Fhal at Clos du Rouge Gorge talked about employing five workers for 8 hectares; in contrast for more robust vines one man would be sufficient for 15 hectares.

Working the coastal terraces with horses

PRUNING

All the older vines are low, stocky bush vines, but the branches must be kept short or they risk being broken by the wind. Gérard Gauby favours *gobelet* pruning, which also lowers the leaf surface, meaning less evaporation, and protects the bunches from the sun. Syrah, with its fragile branches, needs the support of wires, one of the reasons that it does not perform so well in Roussillon. With the shoots trained on wires, the grapes are more liable to burn, so that Syrah needs lots of leaves to provide protection from the sun. Supporting vines *en echalas*, with a single post, helps develop the leaf area and resist the wind. The leaf surface area increases the polyphenols, giving more concentration and colour to balance the impression of alcohol. As Laetitia Pietri-Clara of Domaine Pietri-Géraud observed, the greatest problem for Roussillon's growers is the wind, and planting *en echalas* does help with that, although wind at flowering can still be particularly problematic.

ORGANIC AND BIODYNAMIC VITICULTURE

Organic viticulture is becoming increasingly important in Roussillon, and rightly so when the climate is so favourable to its development. Twenty per cent of the vineyards were certified organic at the last count (see Appendix VI for details), and there are others who work organically but simply have not bothered to register as such. Biodynamic viticulture can be seen as the natural corollary of organic viticulture, but other growers express a preference for HVE, with its emphasis on biodiversity.

Domaine Cazes was a pioneering estate for organic and biodynamic viticulture in the region, beginning the conversion to organic viticulture in 1997. Today it is one of the largest biodynamic wine estates in the whole of France. Domaine de la Rectorie's conversion to organic viticulture took ten years. They relinquished vineyards that were either too difficult to work or did not produce good quality fruit. They use no weedkiller, but all the hand labour inevitably has an effect on prices. They *déracinent* after the harvest, cutting the surface roots of the young vines, to encourage the other roots to dig deeper. Dead vines are replaced, grafted in the vineyard, and they are considering grafting from their own wood.

Eric Monné talked of a new wave of wine growers working organically, making wines with better acidity levels, from the old varieties that might be picked slightly unripe. Acidity was lost in the quest for Vin Doux Naturel, but these days organic viticulture brings out the acidity, making for a highly desirable freshness in both the red and white wines. Pierre Boudau of Domaine Boudau also discussed the development of organic viticulture in the region. In the early years of this century came the pioneers, along with opportunists, who saw it as a marketing opportunity, he noted. But gradually, over the last five or six years, pressure from customers has started to mount, so that the organic label will become a standard. A wine grower who is not working organically will be seen as someone who simply has not understood how to evolve and develop. Jean-Marc Lafage, of Domaine Lafage, would agree: organic is becoming a brand, so that the market demands organic.

For many wine growers, their organic farming has progressed to biodynamic viticulture. Wendy Wilson talks about biodynamics with passion. She is convinced that it leads to more balanced soils, and therefore more balanced wines. It very much depends on observation, you must notice what the vineyards need, she explains, but one must also be practical; there are not enough 'root days' in the calendar for her to prune all her vines on the most appropriate day and she does not view the various biodynamic preparations as precise recipes.

Gérard Gauby complained about the effects of weedkiller and how its use compacts the soil but also noted benefits to organic farming that stretch beyond the vineyards. For example, pesticides had polluted the fresh water spring of the village of Calce, so that it had to be closed. It took ten years to make the water potable again. The well is 8 metres deep and now provides water for the 90 inhabitants of the village, making Calce one of the very few villages to have its own water source.

Domaine Gardiés has been organic for 15 years, and they are now practising biodynamics, but prefer not to be registered, with Victor Gardiés making the pertinent observation that the official *cahier des charges* for biodynamics is adapted to the viticulture of the north, not the south. For example, they do not need to work on improving photosynthesis in the south. Initially organic was a pejorative term and was not taken seriously. However, it is not complicated in Roussillon, where normally the wind keeps everything dry and disease free, so there is no reason why it should not be the norm here, Victor points out. The overall feeling these days is that if you are a responsible wine grower who cares about your land, you must be organic, or working towards it. Otherwise you will lack credibility. An occasional and unusually wet spring, such as in 2020, can, however, cause some serious problems with mildew for organic growers. Alain Razungles had never used insecticides or pesticides so when he converted to organic viticulture his most pressing concern was controlling the grass and weeds. Olivier Pithon likes to work the soil as little as possible and focuses on protecting the vines from excesses of heat and sunlight.

There are many wine growers who in effect work organically, without being certified, as that is what they have always done, without questioning their methods. For them there is no reason not to be organic. However, Boris Kovač at Domaine de l'Agly admitted that he gave up on organic viticulture in 2017, as he needed to save grapes affected by grey rot. His criticism of organic viticulture is that it requires increased tractor use, which compacts the soil, in turn killing off natural flora. There is also the unresolved problem of copper, which is permitted in organic treatments but which carries toxicity risks and persists in the soil. Boris prefers HVE, with its greater emphasis on biodiversity, working to maintain the surrounding *garrigues* and protect the trees. Estates such Mas Lavail are working on various levels of HVE, sometimes in transition towards organic viticulture.

The small cooperative at Cases-de-Pène, which currently practises *lutte raisonnée* or sustainable viticulture, is also considering HVE en route to organic viticulture. They work with the Groupement d'Intérêt Economique Environnemental, analysing the soil, with particular regard to organic matter, to ensure that they use the appropriate fertilizers rather than applying them indiscriminately. For instance, they found that phosphorus was lacking in some of their vineyards, which they were then able to remedy. The Rivesaltes cooperative, Arnaud de Villeneuve, is working with Vignerons Engagés (previously Vignerons en Développement Durable), an organization which concentrates on sustainability, considering not only biodiversity and the environment, but also the social and economic aspects of wine production. It accounts for about 5 per cent of the vineyards of France and is comparable to HVE.

DISEASES AND OTHER PROBLEMS

With the prevalence of drying winds acting as an effective preventative measure, diseases are less of a problem in Roussillon than in most other French vineyards. As Emmanuel Cazes observed, Collioure is the vineyard requiring the least treatment in the whole of France, thanks to the drying wind. Mildew can be a problem, but is the exception rather than the rule, again thanks to the wind. It is less of a problem for vineyards by the sea, due to the sea breezes.

Flavescence dorée has been a problem in the past, and was treated comprehensively, with helicopters systematically spraying all the vineyards in Collioure and Banyuls – there were fewer organic wine growers at the time – as the only practical method in such difficult terrain. Sadly, it is becoming a problem again, particularly at the western end of the Agly Valley around Caudiès. As Wendy Wilson observed, the organic vineyards are more susceptible as they provide a more welcoming habitat to the vector, 'an innocent cicadelle'. There are sprays, with which wine growers are obliged by law to treat their vines three times a year, with organic wine growers using an extract of pyrethrin. The only really effective remedy is to pull up infected vines. Essentially its control demands a lot of observation in the vineyard – there is talk of using drones to monitor the plants. It is often the less well cared for vineyards that are most susceptible. The Conseil Interprofessionel des Vins du Roussillon (CIVR) reckons that just two per cent of vineyards are affected.

Serge Baux at Mas Baux is trialling a new technique to counteract *vers de la grappe*. The trouble with green insecticides is that they also harm the beneficial insects, so this new system works by using pheromone traps to provoke sexual confusion, placing them on every third vine. A new pest has also been discovered, a moth called cryptoblabes that lays its eggs inside bunches of grapes. The eggs are so tiny that you cannot see them with the naked eye, but a Spanish laboratory has developed a pheromone to combat them, and this has just been approved in France. However, Serge observed that in 2019, with two severe heatwaves, there were fewer insects as well as fewer birds. Consequently he did not need to use any insecticides, and very little copper and sulphur.

For Jean Boucabeille of Domaine Boucabeille, it is the wild boars that are the worst plague of all, even worse than climate change. Electric fences are the only feasible solution, but they are complicated and expensive. Philippe Bourrier of Château de l'Ou did not produce any Chardonnay at all from his vineyard in Caudiès in 2017 as the wild boar had consumed the lot. In 2020 there were even more wild boars than normal, as Covid-19 had curtailed the hunting season. Bastien Baillot of La Bancale considers the deer to be even worse, for they eat the young vine leaves, which will ultimately kill the vine.

DROUGHT, IRRIGATION AND CLIMATE CHANGE

Climate change is having an impact in Roussillon, as it is on vineyards everywhere, but the problems that it presents are being addressed by the more visionary wine growers. Prevailing drought conditions have forced change for some of the region's producers. Domaine de la Rectorie's owners are looking for cooler sites, as they have given up the vines that suffered perpetually from drought. Vines suffer more in clay and limestone sites than in vineyards in the region that are on granite, which retains water. Michel Berta of Domaine Berta-Maillol observed that the frequency of drought is increasing: he would irrigate if it would save the vines. However, water availability may become a problem, in which case is it reasonable to use water for vines? And while irrigation may be a solution for vines grown on the plain east of Estagel it is too difficult to irrigate vineyards on slopes. Boris Kovač says he would like to irrigate, but it is simply too expensive.

At Château de Jau Simon Dauré is lucky, in that he has both access to water and flatter vineyards and has been able to irrigate his young vines during a considerable replanting programme. He is working towards both organic viticulture and HVE, and has bee hives on his vineyards and sheep grazing in the winter. For him, 'the climate is a trump card for organic viticulture; with wind and sun, and no neighbours who might pollute.' However he too acknowledges the problems of drought: 'You often need to irrigate just before the harvest so that the grapes can ripen properly without any stress,' he says. Julien Jeannin at Mas Crémat has irrigated since 2019 to alleviate the stress in the vines, from which Syrah in particular suffers. Again, he is lucky that he has water, and relatively flat vineyards, compared to some, on broad terraces.

Heat stress may also be a consequence of global warming. Pierre Escudié at Domaine de Nidolères on the plain outside the village of Tresserre is addressing the problem with solar panels. He realized some time ago the important effect trees can have on a vineyard, and was planting trees 30 years ago, for biodiversity, so concerned was he about global warming. He noticed that a vineyard with a row of apricot trees for every three rows of vines produced wine that was less alcoholic and less tannic. Plants need shade, at least three hours per day, longer in hotter spells, allowing them to rest. He found a company in Lyon, Sun'Air, which installed five hectares of solar panels over newly planted vineyards of Grenache Blanc, Marselan and Chardonnay, which will go into production in 2021. The adjacent vineyard acts as a control, for comparison. The panels are controlled by a computer in Lyon, and change position to alter the impact of the shade. This way Pierre thinks he can slow down and regulate the ripening as well as reducing moisture evaporation from the soil. He is certain that *un coup de chaleur*, a spike of heat, destroys something in the grapes, and makes for a lack of balance in the flavours. Drip irrigation would be too expensive, and much of the water would simply evaporate. Naturally the solar panels also go towards reducing the vineyard's electricity bills and its reliance on fossil fuels. Incidentally Pierre's wife, Martine, is a talented cook, and the restaurant at Domaine de Nidolères is highly recommended.

There is no doubt that small yields caused by drought have a serious impact on the economy of the region (see page 73). And there is no doubt that the yields are tiny. Jean Boucabeille talked of an average of 10–25 hectolitres per hectare, well below the permitted yields for the appellations, which would give you between 13,000 and 33,000 bottles per hectare. Yields are affected intially by pruning, but ultimately, they are determined by rainfall. The average rainfall figures that I was given by Guilhem Soulignac for his vineyards at Domaine Riberach certainly make for disquieting reading. The figures are for the total rainfall from the end of harvest to the following September. When we take into account that the average annual rainfall is in theory 500–600 millimetres, it is clear that both 2015 and 2016 suffered from drought.

2015 250 millimetres
2016 200 millimetres
2017 480 millimetres
2018 1,400 millimetres[2]
2019 700 millimetres

HARVEST DATES

An obvious consequence of climate change is the date of the harvest. Robert Pouderoux observed that in the 50 years before the heatwave of 2003, no harvest had ever started in August. It usually began between 5 and 10 September. However, since 2003, the harvest has always started in August, anytime between the fifteenth and twentieth of the month, notably for Muscat à Petit Grains and Syrah. Marie-Pierre Piquemal of Domaine Piquemal, for example, began her harvest on 10 August in 2016; and on 20 August in 2018. Until 2003, the later-ripening Muscat d'Alexandrie was never picked in September, but always in the first week of October. In contrast, these days it is always picked in the

2 The autumn of 2018 was especially wet, skewing the rainfall picture, but there was very little rain in May and June, apart from three good storms.

last week of September, and never in October. In the 1960s the harvest could begin as late as 10 October; nowadays it has nearly always finished by that date. The difference is enormous. The problem is not necessarily hotter summers, but the lack of rain. Where have the storms of July and August gone?

Hand harvesting at Domaine de la Rectorie

Decisions regarding the picking date are important and vary widely. To retain tension in his wines Cyril Fhal picks at the beginning of the ripening, when the grapes have reached a potential 12–12.5% abv. But with such a diversity of grape varieties and vineyard sites, harvests tend to be quite long and drawn out, starting in mid- to late-August and lasting until early or even mid-October. The harvest usually begins with Muscat à Petit Grains from warmer vineyards nearer the coast, and for red grapes Syrah is normally first, with the harvest finishing with the later ripening varieties, notably Carignan and Mourvèdre.

Hand harvesting is widely practised as mechanical harvesting is not usually feasible, given the narrow rows and steep terraces in many of the vineyards. There is usually no problem finding pickers. Generally, they are people who work in the fruit orchards and harvest other crops, often coming from Spain or Croatia. However, Simon Dauré joked that while it is not politically correct to enthuse about mechanical harvesters, they have improved greatly and now work well. At Mas Crémat half the vines are handpicked, and half mechanically harvested, which has the advantage that you can pick very early in the morning and avoid the heat of the day.

I would like to give the last word here to Pierre Boudau, who commented that the vineyard of Roussillon is still young, by which he meant Roussillon vineyards in their modern incarnation. He observed that Gérard Gauby, who is widely seen as a pioneer, has only done 35 vintages. Pierre himself has not even achieved 20, with a first vintage in 1992: 'We are still in the phase of research, constructing the vineyard,' he says. 'As yet there are no certainties and we must be aware of and beware the phenomenon of fashion.' The wonderful thing about the vineyards of Roussillon is the mosaic of varying plots, with so many different soils, aspects and altitudes, making a veritable artist's palette for the blending of wines in the cellar. The task and talent of the wine grower is to interpret these various terroirs.

4

WINEMAKING: VIN SEC AND VIN DOUX NATUREL

In recent years, winemaking in Roussillon has evolved and improved, as it has all over the south of France. For table wine the improvements have been in all aspects and no particular one stands out. One significant difference between this region and other French winemaking regions is that many of the wine growers in Roussillon make both fortified wine and table wine, both Vin Doux Naturel and vin sec, and from the same vineyards. However, the wine growers of Roussillon have relatively limited experience of making vins secs. Until the last 20 years or so, Vin Doux Naturel was a much more important aspect of their work; it is really only in this century that winemaking has focused on vin sec, and Roussillon winemakers have had to learn how to make vin sec, almost from scratch. As Mickaël Sire of Domaine des Schistes noted, in the 1990s they copied existing models, and then in the 2000s began to develop their own style.

These days the search for freshness, in all three colours, is a significant factor in determining winemaking decisions. On their way out are the heavily oaked, over-extracted reds, along with whites that lack acidity. Indeed, when you consider that they come from some of the hottest vineyards of France, it is astonishing how fresh and lively the white wines of Roussillon are.

Let us consider the vins secs first. As discussed in the chapter on viticulture, the key grape varieties for white wine are Macabeo, Grenache Blanc and Grenache Gris. You might find Tourbat, otherwise known as Malvoisie du Roussillon. Vermentino, Roussanne and Marsanne make appearances, while the international varieties of Chardonnay, Sauvignon Blanc and Viognier are very much less widespread than in the Languedoc. You will also find Carignan Blanc, and even Carignan Gris, and of course, Muscat à Petits Grains and Muscat d'Alexandrie, for both Vin Doux Naturel and, increasingly, Muscat Sec, which was developed when the market for Muscat de Rivesaltes began to fail. However, the overriding feeling is that most wine growers prefer to work with other varieties, such Grenache Gris, Carignan Blanc, Macabeo and Vermentino, which they see as more interesting, and also more remunerative.

As for red wine, Grenache Noir has long been the mainstay of the vineyards of Roussillon, often *complanté* with Grenache Gris, Grenache Blanc and Carignan. Syrah and Mourvèdre are more recent arrivals; there is Counoise too, but very little Cinsault. The likes of Cabernet Sauvignon and Merlot do not really contribute to the flavours of Roussillon. The emblematic variety is undoubtedly Grenache, in all three colours.

WHITE WINE AND ROSÉ

Every effort is made to have cool grapes arriving at the press: some producers have refrigerated containers to cool the grapes overnight if necessary. Care is taken over any potential oxidation and cellars are well-equipped for temperature control. The winemaker decides whether or not they would like a malolactic fermentation. For white wine, *bâtonnage* is fairly rare as it can make the wine too heavy, though the wines may well spend time on their lees. Some skin contact may also be an option; Michel Berta likes it for Grenache Blanc, but not Grenache Gris.

Boris Kovač told me in detail about his white winemaking. The grapes are handpicked in the early morning, and collected in small *cagettes* or boxes. In the cellar, he has a sorting table, and fills the barrels gently as the fermentation starts. There is regular *bâtonnage*, initially four or five times a week, dropping gradually to once or twice a week, and then once or twice a month. Boris enthused about blending: each variety is vinified separately, in small barrels or tronconic barrels, with a manual *pigeage*. He is interested in the tannins, tasting the pips at the beginning of the harvest, and considers that the oak is better integrated if the wine is fermented in barrel. Barrel ageing for white wine gives body and weight.

For his white wines, Pierre Gaillard said firmly, there is no new oak. He favours a long, slow fermentation, working on the lees, to retain minerality. Oxygen is the enemy. However, Macabeo can be very reductive, influenced by the atmospheric pressure with sea breezes, causing the lees to move. The strong reductive impact of the lees is balanced out with Grenache, which is oxidative, and the strong acidity of Carignan Gris.

Most rosé wines are made by pressing the grapes, either immediately or after a very short period of skin contact, and then the juice is treated in much the same way as for a white wine. Cathy Sisqueille of Château de Rey however thinks that more intense rosés are more interesting: *saignée* rosés retain their intensity. On reflection rosé does not feature greatly in the repertoire of most Roussillon wine growers. William Jonquères d'Oriola at Château de Corneilla takes his rosés more seriously than most. The grapes are picked at night, pressed immediately and allowed a malolactic fermentation, which adds some weight. For his most serious wine, Cavalcade, fermentation takes place in oak, for a rounded mouthfeel. Jonathan Hesford at Domaine Treloar makes a pure Mourvèdre rosé, a food rosé, pressed and given a cool fermentation in a stainless steel tank, and Dominique Grénot at Mas Llossanes has a pure Cinsault rosé.

RED WINE

There are probably as many opinions about winemaking as there are wine growers. Here some of the voices of Roussillon add to the discussion.

The use of sorting tables, allowing for an opportunity to remove any inferior grapes, is more widespread in the region than ever before. The next consideration is a pre-fermentation maceration, which is good for Grenache, adding colour and concentration, but does nothing for Carignan. Jacques Castany at Domaine de l'Edre favours a cool pre-fermentation maceration which helps retain the fruit, followed by a shorter maceration during fermentation in order to avoid any excess of tannin. Pierre and Elise Gaillard give their Collioure a pre-fermentation maceration for about five days at 8°C. The grapes are all destemmed to avoid any bitterness and care must be taken over the drying aspect of the tannins. It is all a question of balance. Robert Pouderoux in Maury said the key to avoiding too much tannin, and too much extraction, is to work gently, with little movement of the wine, and to rack gently, with the juice protected by carbon dioxide. Gravity-run cellars to minimize pumping are best. Charles Perez also wants to avoid too much extraction so favours a long pre-fermentation maceration. During the fermentation he does some *délestage*, but very little *remontage* and no *pigeage*. If the skins are fragile, which those of Grenache can be, they are spoilt by *pigeage*. He wants velvet tannins, which put the fruit *en valeur* i.e. emphasize the fruit. Some people *égrappe* (destem) – *égrappoirs* are now quite sophisticated, and can eliminate any dry berries – others prefer whole bunches or a mixture of both. At Mas Baux, Serge Baux has a mechanical system of *pigeage* for each vat, which dampens the grapes with a wave of juice from the bottom of the vat over the cap, twice a day, avoiding the use of a pump. He does not like *remontage*.

Carbonic maceration, the fermentation of uncrushed bunches of grapes in a vat already filled with carbon dioxide, is much less widespread than it was when the appellations for vins secs were created. Carignan responds well to the technique as it softens the sometimes rustic tannins of that variety, and so this method continues to be used for this variety. The technique, or a semi-carbonic maceration, without the addition of any extra carbon dioxide to the vat, is also popular amongst those making natural wine as the carbon dioxide protects the wine against oxidation. It can also make for earlier drinking wines. In contrast, Cédric Haague observed that he has used the technique just once in 20 years.

Gérard Gauby's winemaking has changed over the years. Originally he fermented whole bunches, then he bought a destemmer, and now he and his son, Lionel, have returned to whole bunches, which they feel give more elegance. When he started out, Gérard used selected yeasts, but these have been abandoned in favour of indigenous yeast. Their sulphur levels are as low as possible, as they are with all the thoughtful winemakers of the region. Mickaël Sire has made changes since taking over the winemaking at the family estate of Domaine des Schistes ten years ago. He is now looking for fresher wines with less marked élevage, using larger containers. Jean-Philippe Padié uses whole bunches in his winemaking, describing it as an infusion with no forced extraction. A third of the vat is filled with juice, to which he adds whole bunches, with no *pigeage* or *remontage*, just an infusion for five days. The pressed juice and the free-run juice are then mixed together for the fermentation to finish. He explained that he has used this technique since 2014, having developed it when he was lacking vat space for some Carignan and put it in the press, adding the juice to a vat where there was room. Séverin Barioz at Face B, who acquired practical experience working with Jean-Philippe Padié, explained the idea of a *millefeuille*, with layers of *égrappé* and non-*égrappé* grapes, and about one-third immediately pressed juice, which means there is no need for *pigeage*. You could call it an infusion, without any sulphur.

The length of the maceration during fermentation is another important consideration. For Victor Gardiés, it could be 8, 12 or even 18 days, depending on the characteristics of the vintage. However, all would agree that there is a need to work gently, *travailler doucement*, and avoid excessive extraction. Alcohol levels, though, can be an issue. The wines of Roussillon have always been strong, hence their suitability for Vin Doux Naturel. Vin sec can reach 18% abv, with the average being 15% abv, but there is definitely a move towards lower alcohol levels, determined by work in the vineyard as well as in the cellar. Oliver Pithon wants freshness combined with generosity, so he does not pick too early, saying that Grenache at 12–12.5% abv is not for him.

Sélection parcellaire or site selection is a growing trend, especially when newer streamlined cellars allow for more vats and smaller vats. Blending too is an essential part of the winemaking of Roussillon; most estates have vineyards in several villages, and consequently different interpretations of the same grape variety from different soils. The appellations insist on blends of grape varieties so *monocépages* are only found in IGP Côtes Catalanes or Vin de France from the region.

ELEVAGE

The use of oak for élevage is one of the main variables, for all three colours, though inevitably it is used less for rosé. Remember that for many estates making vin sec is a relatively recent activity, and as Mickaël Sire observed, Roussillon only has about 20 years' experience of élevage.

Initially small Bordeaux barriques were favoured, but these days the trend is towards demi-muids or even larger foudres, and often older barrels rather than new ones. Victor Gardiés spoke eloquently about the use of oak, explaining that he does not want hard tannins and structure but instead looks for balance and freshness. He takes care over extractions, preferring an infusion of juice and grapes rather than any violent *pigeage* or *remontage*. He does not crush the berries, but retains some whole bunches and berries as he wants what he called the *côté éclatante*, brightness, and a balance of black and red fruit. He has experimented with barrels of different sizes, including 6- and 20-hectolitre foudres. He does not like the taste of wood; a wine needs élevage, he says, but it should not be noticeable. He wants wines that will retain their tension and age well, leaving a finish that makes you salivate. We must not forget that wine is meant to be drunk, but we don't know how to make *des vins immédiats*, for early drinking, he notes.

Jean-Roger Calvet explained how his élevage has changed over the years. He began with barriques, but found he lost the profile of the wine. His barriques are now used for Vin Doux Naturel, and for vins secs he has gradually changed his approach, first trying 600-litre demi-muids, and then small foudres. In 2009 he bought some 15-hectolitre foudres, and then in 2019 two 25-hectolitre foudres. He considers that foudres give more *finesse* and elegance. Château de Rey is unusual in having some barrels made half from oak and half from acacia. They like the taut note of acacia, but there is no tradition for it in Roussillon.

There is also a lot of élevage in concrete vats, while stainless steel tanks are used to retain freshness and fruit. At Mas Amiel you can find just about every container you might imagine: cement and stainless steel vats, tronconic wooden vats, small barrels, amphorae and tiny horizontal cement vats. There are some 6-hectolitre concrete vats, which they call sputniks, due to their curious shape. They are undecided on whether the shape has any impact or not. They also consider the porosity of the vessel, with oak the most porous of all. The amphorae are useful for smaller cuvées of site selections.

Vinification intégrale

In Roussillon you will also encounter some *vinification intégrale*, entailing a fermentation in a small and possibly new barrel, such as for the top wine of Domaine de l'Edre. Likewise, Volte Face at Domaine la Toupie is vinified in new wood, followed by élevage for nine months, making for some fairly overpowering flavours. At Château de l'Ou, Séverine Bourrier makes two wines in this way, both Syrah from different vineyards, with intensely chocolatey fruit and more obvious oak and weight.

Jérôme Collas from Domaine la Toupie is experimenting with élevage in plastic eggs, which apparently have a similar porosity to a two- or three-year-old barrel. He stirs the juice occasionally and so far has found that it refines the tannins of the Grenache and that the wine is ready earlier, with more rounded tannins. He sees the egg as an alternative to wood, especially for Grenache, and is planning to buy some more. He might try an amphora, which he expects to enhance the freshness of his white wine.

Amphorae and eggs

Philippe Bourrier at Château de l'Ou talked cheerfully about amphorae, *le jarre du fainéant*, the amphora of the lazy, as the lees stay in suspension with a gentle movement or *bâtonnage* without any need to touch them. An amphora gives freshness to both red and white wine. The thickness of both an amphora and an egg is important, with 3–5 centimetres giving the right exchange of air. Philippe also has an egg made from oak; as with a concrete egg, the wine moves. He is one of only two French wine producers to have wooden eggs; the other being Château Pontet-Canet in the Médoc. Dimitri Glipa at Mas Mudigliza enthused about eggs, which he uses for his white wine, explaining how they put the fine lees, but not the *grosses* or larger lees, into suspension and without any contact with oxygen. With amphorae, there is more oxygenation than with wood, but you lose less wine. At Clos St Sébastien in Banyuls, Pierre Girault is experimenting with amphorae for white wine, for which the first vintage was 2019, while for his red wine he uses foudres or barriques. He also has some lightly toasted Stockinger barrels for his white wine.

ORANGE WINE

There is a growing interest in orange wine, not just in Roussillon, but in many other regions, both in France and elsewhere in Europe and beyond. Quite simply, orange wine entails the fermentation of white grapes on their skins, in just the same way as you ferment red wine on the grape skins. The juice takes on colour from the grapes and may indeed acquire some orange hues, and the juice tends to be more tannic. Jean-François Deu at Domaine du Traginer made his first orange wine in 2017, mainly from Grenache Blanc and Grenache Gris, giving it a four-week maceration and élevage for two years in demi-muids. He was inspired by the wines of Stanko Radikon in Friuli and observed that it is easier to make orange than white wine, as sometimes there is a problem with finishing the fermentation. As well as his traditional, fresh Muscat de Rivesaltes, Mickaël Sire makes an orange Muscat, giving it a ten-day maceration. He calls it a Muscat 'with peps', and it has some characterful bitterness and refreshing orange notes.

Justin Howard-Sneyd and Jean-Marc Lafage work together to make Taronja, which means orange in Catalan. It is a blend, mainly of

Grenache Gris and Grenache Blanc, with a little Muscat and even less Viognier. In 2018 there were four barrel fermentations, two with whole bunches, one carbonic maceration and one with destemmed grapes. The fermentation took place in open-top 500-litre barrels, after which the juice was left to macerate on the skins and stalks for two weeks. The wine was protected from oxidation with dry ice and then aged in neutral barriques for eight months. There was a light filtration, but no fining before bottling.

Tom Lubbe at Domaine Matassa is another enthusiast for orange wines, going so far as to observe that his palate has changed and he now gets rather bored drinking white wines. Virtually all his white grapes are fermented on the skins, with Cuvée Marguerite, a blend of both Muscats and Macabeo spending between 14 and 30 days on the skins, making for more subtle Muscat flavours, if that is possible. The orange wines are either Côtes Catalanes or Vin de France, but never an appellation.

I greatly enjoyed the orange wines that came my way when researching this book. The best have a refreshing character, enhanced by more tannins than in a conventional white wine. With the growing worldwide interest in the style, I would expect them to become a small but significant part of the repertoire of Roussillon amongst its more adventurous wine growers.

NATURAL WINE OR VIN NATURE

While most serious winemakers will try to limit their use of *intrants* or additives, in particular sulphur, Roussillon has attracted a growing number of winemakers who would describe themselves as natural winemakers. Until very recently there was no *cahier des charges* for *vin nature* but in February 2020, the Syndicat de Défence des Vins Naturels was created. It drew up a list of requirements for producers wishing to use the label 'Vin Méthode Nature', including organic viticulture, hand harvesting, no additives, no filtration and use of only natural yeasts. Hygiene in the cellar is paramount. Two different logos are available for use by members, depending on the sulphur content of the wines: one for wines with no added sulphur and the other where sulphur has been used, usually at bottling, but to a level of less than 30 milligrams per litre.

The village of Latour-de-France has attracted a growing nucleus of natural winemakers. Raymond Manchon of Domaine Raymond

Manchon, a natural winemaker in Estagel, offered an explanation of this snowball effect. The Agly Valley is definitely drawing attention, but another significant factor is the financial aspect. There are vines for sale in interesting terroirs that are not too expensive. Natural wines can usually be sold sooner than more conventional wines, which might require a longer élevage, and that too helps those with limited financial means.

Raymond himself enjoys the notion of liberty with natural wine, escaping from the *cahier des charges*, but notes that with less stable wines natural winemakers face the permanent risk of a problem, which keeps them on their toes. The standard practice has become a lower alcoholic degree, a short maceration and an early bottling. Raymond, on the other hand, prefers a three- if not four-week maceration, and even on one occasion a six-week maceration, determined by the tannins. He does not want to impose a technical process.

Talking to natural wine producers, you cannot help but be aware of their philosophical motivation, and for many it is a way of life. Matthieu and Vanessa Courtay of Le Mas de la Lune see it as the natural corollary to biodynamic viticulture. For Cyril Fhal, who is a meticulous winemaker, natural winemaking is certainly not a question of doing nothing. 'We do a lot so that the wine develops on its own,' he says. Ideally, he uses no sulphur, but 'it is not a religion'.

Philippe Gard makes a *vin nature* using whole bunches, with a short maceration rather than a carbonic maceration, and an early bottling. The yeasts are indigenous and there is no added sulphur, with a label telling the consumer to decant. The result is fresh, with perfumed red fruit.

Several other estates that usually make more conventional wines include a low sulphur or even no added sulphur wine in their repertoire, such as Simon Dauré at Château de Jau with Vin Nature, 0%, although he admits that he is not a fan of *vin nature*. William Jonquères d'Oriola has begun to produce L'Indigène, Sans Sulfites Ajoutés, fermenting whole bunches in a carbonic maceration. At Mas Janeil they make two Côtes du Roussillon Villages, one with sulphur and one without, for which they may or may not use the indigenous yeast. At Château de l'Ou, Séverine Bourrier makes Esprit Libre with no sulphur; the first production was in 2018. Her aim was fruit, and she gave the destemmed grapes a pre-fermentation maceration at 4°C, gradually increasing the temperature, and adding cultured yeast. The same wine, made with some added sulphur and riper grapes, had more weight. There are many more examples.

A recent film, *Wine Calling*, directed by Bruno Sauvard, focuses on a group of natural winemakers, Jean-François Nicq at Domaine les Foulards Rouges, Stéphane Moran at Domaine Léonine and Laurence Manya at Domaine Yoyo, among others, who talk about their winemaking, and more importantly the philosophy behind their winemaking, with immense commitment. There is more risk but more freshness; it is a choice that they make, a way of life. But an observation by Laurence Manya really struck me; that it was the use of chemicals 50 years ago that saved the vineyards of Roussillon. Without chemicals in such difficult terrain, these vineyards would have simply disappeared as they would have been impossible to maintain. Boris Kovač is sceptical about natural wines, admitting that he likes to control things, and to respect certain rules when he makes a wine, much as he would when driving a car. Vincent Parcé of Domaine Augustin suggested the need for a best before date for these wines, as without sulphur wine is a perishable product and declines too soon.

One must be wary of fashions. But there is no doubt that the climatic conditions of Roussillon, allowing for reduced use of sulphur, are favourable to both organic viticulture and natural winemaking. Organic viticulture is growing in importance and natural winemaking, which begins with organic viticulture, is certainly here to stay. Sadly, some of the wines may have faults, as they are less stable than a more conventional wine, but natural winemaking should not be used an excuse for a fault, and the best are undeniably delicious.

VINS DOUX NATURELS – REDUCTIVE AND OXIDATIVE, MUSCAT AND RANCIO

Vin Doux Naturel covers several different styles of wine. All are fortified, and the differences come with the ageing process, or indeed the lack of élevage. The oxidative wines are traditional, whereas the reductive winemaking for Rimage and Grenat is more modern; they are for the new generation, with freshness and fruit. As Victor Gardiés put it, vin sec is more demanding, whereas with Vin Doux Naturel, the sugar camouflages any faults. For him, 'it is a wine for *fainéants*', i.e. for the lazy.

Mutage

The first thing to decide is the right moment to add the alcohol, the grape spirit, and stop the fermentation. There is a broad band of residual sugar allowed, anything between 75 and 130 grams per litre, but it is usually 80–100 grams per litre. If there is too much sugar, the wine will be heavy. Daniel Laffite at Domaine des Soulanes talked about sliding the tannins behind the sugar, so that you have ripe fruit balanced by a streak of tannin. For his Grenat he mutes at 87 grams per litre, observing that at 10 grams more of sugar per litre the tannins are masked. If you mute three days later, you will need less alcohol and the wine will be drier. It is all a question of balancing sugar and alcohol. The tannic streak can be quite refreshing. The volume of alcohol depends upon how ripe the grapes are. Ripe grapes need less alcohol, which usually accounts for about 8 to 10 per cent of the volume. Philippe Gard at Coume del Mas wants a dry style for his Rimage, with some tannin, and some barrel ageing for a few months. The wine is muted at 80 grams per litre, with the tannin balancing the sugar.

The other decision is whether to mute *sur juice*, or *sur grains*, before or after the juice is racked off the skins. It is more difficult to mute *sur grains* during the fermentation – it is harder to gauge the right time since you do not really know how much sugar there is in the skins. For this reason Alain Razungles described *mutage sur grains* as Russian roulette. The wine may be sweeter or drier than desired, so a solution is to make several batches and blend them. The advantage of *mutage sur grains* is that it gives more concentration and ageing potential, and makes for better integrated alcohol. In the past another factor in deciding when to mute was cost. It used to be cheaper to mute *sur jus*, as you had to pay the tax on the fortifying alcohol when you bought it. Now you pay the tax when the bottle is sold, with the result that today more people mute *sur grains*.

Ageing

Rimage and Grenat can have wonderfully ripe vibrant fruit, but there is no doubt that the true originality of Roussillon comes from the wines that have been aged for a number of years. For Hors d'Age that is at least five years, but very often much longer, so that the wines develop the most delicious *rancio* flavours. These are among the vinous treasures of France. *Bonbonnes,* foudres and barriques are all used for ageing Vin Doux Naturel, both inside and outside. The barrels for Vin Doux Naturel must be subject to the elements and to the changing seasons, with extremes of temperature. They are often stored in an attic, as they are at the cooperative in Fourques. The barrels are never topped up and over about 45 years with a 3 to 5 per cent evaporation they would be empty. The evolution is not linear; the winemakers taste regularly and decide what to bottle and what to continue ageing.

Philippe Gard has his barrels in his small cellar, on top of the vats, right under the roof. The barrels are not topped up as evaporation – the 'angels' share' – is a significant part of the process, and enhances the flavours of *rancio*. Barrels are often left outside so that they undergo a thermal shock, while those inside allow for less evaporation with a smaller angels' share. Bernard Rouby, a previous president of the cooperative of Maury, observed that in Maury, the angels are thirstier, as the atmosphere is drier than in Banyuls; the difference between mountain and sea. The Banyuls cooperative, Terres des Templiers sprays its barrels with water to avoid excessive evaporation. However, the angels' share is important; as the evaporation concentrates the wine. You can put Vin Doux Naturel in a barrel and simply forget about it, as many have.

Ageing in foudres at Mas Amiel

In old fashioned cellars like Château de Rey you will see enormous old foudres. They were the traditional vessel for Vin Doux Naturel. At Château de Rey, they are empty, but those at Mas Amiel are still used, as are those at Domaine Cazes and Terres des Templiers. There is more evaporation with a foudre as the surface area is so much larger.

Glass *bonbonnes* add another dimension, as they do not allow for any evaporation. However, the sun can also kill flavour, and two years is usually considered the maximum time for a wine to spend in a *bonbonne*. Barrels are gentler, with the wood adding more complexity. Mas Amiel still has a wonderful *parc de bonbonnes*, 800 of them, in which the *rancio* wines spend about 12 months, before being aged in foudres.

The solera is another aspect of the ageing process and often an important part of an Hors d'Age – and you will find old soleras in Roussillon. Domaine Madeloc has one begun in 1920 that was forgotten about in 1962 and only rediscovered when the current owners bought the estate in 2002. Olivier Saperas at Domaine Vial Magnères has a solera that his father started in 1968, with seven stages of barrels, which are kept inside. Some wine is bottled every couple of years or so. The essence of a solera depends on the younger wine taking on the characteristics of the older wine. It is a wine without a vintage, but with a minimum age of ten years, and many Hors d'Age are the result of a solera.

Cellars for Vin Doux Naturel are traditionally dry, so that the wine evaporates, leading to higher alcohol, rising from 17% abv to 20% abv, and making for a concentration of flavour. Georges Puig of Domaine Puig-Parahy insisted that you should never clean the barrels, though he might transfer the mother from one barrel into another. Victor Gardiés agreed that you do need a mother, the lees from a previous wine, and again you should never clean the barrel. Your nose will tell you if there is anything amiss.

Old vintages are part of the patrimony. It is not unusual to find wines that are 30 or 40 years old. Domaine de Rancy have made a speciality of their mature vintages, and Bruno Cazes went to great pains to save the stocks of the Cave de l'Etoile, with wines going back to 1947. More modestly, Mas Amiel still has a 1969.

RANCIO SEC

The history of *rancio* wines in Roussillon is as old as winemaking in the region. Pliny refers to wines exposed to rain and sun and all the injuries of the air. In the nineteenth century *vi ranci* was widespread in Catalonia, and was served at a dinner in 1909 alongside Château Margaux and Château d'Yquem. But later in the twentieth century the oenologists waged a war on oxidation and *rancio* was sacrificed on the altar of modern winemaking.

Rancio Sec is quite different from Vin Doux Naturel as it is not muted, but instead is simply left in a barrel that is not topped up, so that it might develop a *flor*, or *voile*, with some deliciously incisive oxidative flavours. Some would say that as a wine it predates the fortified wines. Confusingly the term *rancio* can also apply to an oxidized fortified Vin Doux Naturel, as it describes a certain flavour of oxidation. Unfortified Rancio Sec very nearly disappeared from the repertoire of Roussillon, as it did not conform to any appellation regulations. Fortunately, the Slow Food movement helped change perceptions about Rancio Sec and a small group of wine growers began to make it again. With its revival in fortunes, it has been allowed for IGP Côtes Catalanes since 2012. No one has ever asked for an appellation for Rancio Sec, so if the wine is not Côtes Catalanes, it is a Vin de France. The exemplary producer is Domaine de Rancy, with three different Rancios Secs, from Macabeo and Carignan, as well as a Syrah, which was made by mistake! They are also amongst the best producers of old Rivesaltes.

Bruno Ribière of Domaine Ferrer-Ribière talked passionately about *rancio*, saying that it is part of the Catalan fibre, and a tradition fundamental to society in the region, dating from a time when many families simply kept a barrel of old wine. A barrel that is not topped up will eventually turn *rancio* in character. They are 'les vins des papis et mamies, un vin d'apéro', which were drunk outside in the villages on summer evenings. They are also great for cooking. Bruno has a *rancio* solera, made from Macabeo, that he began in 2001.

In Roussillon, you will also find wines that are labelled Vin sous Voile, such as those from Domaine de la Rectorie and Domaine Boucabeille. Vin sous Voile has a layer of *flor*, which comes naturally when a wine is aged outside for two or three years. It is not fortified, since they do not want the alcohol to be too high, no more than 14.5% abv. Jean Boucabeille described his method as empirical. You need a sufficient amount of alcohol to avoid the development of any acetic bacteria and he keeps his barrels in the coolest part of the cellar so that the temperature is below 20°C. The *flor* occurs quite naturally. For Jean, the wine starts to become interesting when it is five years old, and becomes even more so at seven years. However, he emphasizes that it is not a *rancio*, but a good oxidative white wine.

MUSCAT DE RIVESALTES

In the past it was more usual to make Muscat de Rivesaltes with both Muscat d'Alexandrie and Muscat à Petits Grains, in equal proportions, but nowadays there is more flexibility. As a result there is more variation in flavour, depending on the proportions of each, or indeed the single variety, used. Muscat à Petits Grains is considered to perform better in the Agly Valley, whereas Muscat d'Alexandrie favours Les Aspres. Julien Jeannin picks his Muscat à Petits Grains at a low level of ripeness, so that it gives a more lemony flavour, and picks the Muscat d'Alexandrie when it is riper, which makes for more body. For him Muscat should have a fresh finish. In contrast, Marie-Pierre Piquemal likes to age a little Muscat each year.

The appellation regulations call for mutage at between 100 and 120 grams per litre of residual sugar, with the alcohol making up 10 per cent of the volume. Twenty years ago, the tendency was for 120, or even 125, grams per litre of residual sugar whereas these days 105–110 grams per litre is generally preferred. With lower sugar, more alcohol is added to produce 15% abv. For Victor Gardiés the balance is very complicated: if you have less sugar, the alcohol can give an impression of sweetness, but if the sugar is too low, the wine can seem heavy. Below 100 grams per litre residual sugar you do not obtain the right balance of sugar and acidity. A Rivesaltes at 45 grams per litre is simply not possible; the sums do not add up.

The mutage of Muscat depends on several factors. Serge Baux said that it ripens early in his vineyards but he has to wait for the *ban des vendanges* before picking, and if it is too ripe by that stage, 16% potential alcohol rather than 15%, the mutage is complicated. Grapes that are too ripe will lead to wine being classed as a *vin de liqueur*, for which the tax is higher. In terms of sugar, there is a big difference between 102 and 105 grams per litre. Serge wants fruit and finesse, otherwise it goes straight to your head, although in French they actually say that it falls to your ankles.

Muscat is so powerful that it can frighten people off with its intense perfume. Pierre Boudau feels that Muscat d' Alexandrie makes for more finesse, and his Muscat de Rivesaltes comes only from Alexandrie. He finds Muscat à Petits Grains is better for dry wine. The cooperative at Baixas always makes a point of having two Muscats, the fresh young Dom Brial and the more mature and complex Château les Pins, which demonstrates the perhaps unexpected ageability of Muscat de Rivesaltes.

So the key to Roussillon is its enormous variety. The *monocépages* can be fabulous in their purity, although the final result may be less complex. However, for Olivier Pithon *assemblage*, or blending, represents 'le côté magique', the magical side of Roussillon.

5

THE BUSINESS OF VIN SEC AND VIN DOUX NATUREL

Roussillon is unusual in that it is one of the rare regions that produces both table wines and fortified wines – the other significant one is the Douro Valley, and in France the appellation of Rasteau in the southern Rhône also includes Vin Doux Naturel, as does Beaumes-de-Venise. Like the Douro Valley, Roussillon established its reputation on fortified wine, but as the market for these has declined, focus in Roussillon has turned to vins secs. Consequently, two themes dominate the wine scene of Roussillon: the successful development of vin sec, and the attempts to reverse the decline in the fortunes of Vin Doux Naturel. The two are linked; for instance, the original success of Banyuls led wine lovers to Collioure.

VIN SEC

Vin sec is a relatively recent development in the region; the first appellations in Roussillon were for Vins Doux Naturels. The first appellation for vin sec did not come until the recognition of Collioure Rouge in 1971, with Côtes du Roussillon and Côtes du Roussillon Villages following in 1977. The main producer of vin sec at the time would have been Les Vignerons Catalans, with a host of village cooperatives providing the wine. Victor Gardiés remembered his father, Jean, making vin sec in the 1980s, with 1994 the first vintage to be actually sold in bottle; his grandfather had only ever made Vin Doux Naturel. The creation of the appellations entailed the planting of Syrah and Mourvèdre, as *cépages améliorateurs*. These days, the introduction of Syrah, in particular, is open to discussion.

The originality of Banyuls and Collioure is the ability to make two completely different styles of wine from the same plot of land. For Philippe Gard, the quality of Collioure stems from that possibility, enabling you to make the fortified Banyuls in the very hot years when a Collioure would be too heavy and alcoholic. However, you have to work even harder to sell Banyuls, he adds. Simply put, people buy more vin sec; you need ten times as many clients to sell the same amount of Vin Doux Naturel.

According to Vincent Parcé, Collioure has not been selling well. The image of Collioure is heavy and powerful, even hard, but, made correctly, it can be fresh and elegant. Collioure Rouge does not have much identity, and flavours can vary quite significantly, depending on which grape variety is dominant. There is also an image of volume, which the new generation of wine growers is trying to reverse. Vincent considers that Collioure Rouge should be seen as the Côte Rôtie of Roussillon. It is potentially larger in size, but the terrain is equally challenging. In contrast, Collioure Blanc is more identifiable, as part of a family of southern white wines. Vincent observed that it is easier to sell Sancerre than Collioure as it is so much better known. All the appellations that make sweet wine also make a dry wine, even Château d'Yquem, with Ygrec. Collioure needs to be explained; the wine growers are not accustomed to going out to sell their wines, and 60–70 per cent of the production is sold locally but, says Vincent, 'we need to take up our pilgrim's staff.' Philippe Gard agreed that when conditions are difficult and an appellation is small you need people with personality who make wines with character, and you need to emphasize something besides technique. He told me that his parents, who came from the Loire Valley, did not drink Vin Doux Naturel; the road from Sancerre to Rivesaltes Ambré is long.

I met Philippe Bourrier of Château de l'Ou when he was president of the CIVR, the body that manages the wines of Roussillon. The position of President alternates between wine grower and négociant and Philippe has now been succeeded by Stéphane Zanella, the director of Vignerons Catalans. I asked Philippe about the aims of the CIVR, to which the answer, in a nutshell, is to develop and improve the reputation of the wines of Roussillon. They want to shift away from sales *en vrac*, for both Vin Doux Naturel and vin sec, and *valoriser*, enhance the value of the wines, concentrating on the region's significant advantages of climate and terroir, which are so favourable to the development of organic viticulture. They also consider carefully the various grape varieties that contribute to the quality of the wines of Roussillon. Philippe mentioned particularly Grenache Gris, but Grenache Noir and Grenache Blanc are also significant, not forgetting Macabeo and Muscat. The Conseil has good contacts with other producers of Grenache, notably from northern Spain, with the vineyards of Catalonia on the southern side of the Pyrenees.

Sales in France are becoming more difficult, so they organize what he called 'missions' overseas, to North America and Asia rather than Europe, and welcome a constant stream of journalists from all over the world. The 1990s saw the development of the export market so that today about 200 wine producers export to some 80 different countries. Exports increased by 40 per cent between 2012 and 2019 and include sales into new markets, such as South Korea and Vietnam. Japan too is working well, and in 2017 China became the largest export market for both Vin Doux Naturel and vin sec. Charles Perez at Mas Bécha has developed a very good market in China and is no longer, as he put it, a slave to the *négoce*. In Europe, Belgium and Germany are important, as is Switzerland, while the UK is twelfth or thirteenth in the pecking order. Russia was important, but the Russian mafia proved problematic so that sales route ended. Explaining the difficult home market Philippe pointed out that Roussillon accounts for just two per cent of the French wine market, equally split between Vin Doux Naturel and vin sec. One significant issue is that 'we are so far from Paris, with small appellations that are virtually unknown in the capital.' Georges Puig dryly observed that Rivesaltes is difficult to pronounce, even for a Parisian. Professional bodies are always open to criticism and many of the smaller growers feel that the CIVR tends to concentrate on the large estates, and that it markets just the appellations and does not push Vin de France from the region. Often the salons and wine fairs are too expensive for a small producer and some producers feel that the CIVR could be more helpful in giving them a presence. A recent marketing initiative, 'les Roussillon sont là!' or in English, 'Reach for Roussillon!' is apparently enjoying some success, though I have to admit that I have been unaware of it in the UK.

Philippe talked about the future of the appellations, which he sees as a mosaic of terroir within a clearly defined amphitheatre. The reputation of the various appellations depends on the wine growers; the appellations are *vins d'assemblage*, blends of grape varieties, and the new wine growers do not always have the right composition in their vineyards, or fall foul of other constraints. For instance, you cannot make Maury if you do not have a cellar in Maury. In contrast, the IGPs, Côtes Catalanes and Pays d'Oc are very broad and, particularly in the case of Pays d'Oc, are not always identified with Roussillon. Philippe enthused about what he called 'la pépite de Roussillon', the nugget, which in fact can be several nuggets. The fundamental question they have to address, is where to put the money for promotion, which is provided by a levy on each bottle. With a membership of 377 independent wine growers and 24 cooperatives (plus 40 wine growers and 4 cooperatives in the Aude that produce Vin Doux Naturel), not to mention all the cooperative members, 34 négociants in the Pyrénées-Orientales and 37 in the Aude, and all the other négociants based elsewhere in France who deal with wine from Roussillon, all with different strategies and objectives, it is far from straightforward. I felt that Philippe was quite a shrewd diplomat. He certainly needed to be.

VIN DOUX NATUREL

The success of Byrrh had the most extraordinary impact on the wealth of the department. As Bernard Rouby observed: 'it created our wealth but it was also a trap.' In its heyday Byrrh was phenomenally successful, but then came to be identified with cheap industrial brands of Vin Doux Naturel. The simplest wines were seen as the *apéro des pauvres*, a poor man's aperitif, without any image and the négociants never tried to change that.

The real crisis for Vin Doux Naturel came during the 1980s, when the production of the cooperative in Maury plummeted from an annual 45,000 hectolitres, 5,000 of which were vin sec, to 20,000 hectolitres, 4,000 of which were Vin Doux Naturel. It used to be easier to sell wine *en vrac* than in bottle, but now the bulk market for Vin Doux Naturel has disappeared.

In the past, Vin Doux Naturel was important for its sugar, but that is no longer the case. Georges Puig explained that 300 years ago, people did not have access to sugar, and that was so until the beginning of the twentieth century. When Vin Doux Naturel fell from favour, people switched to drinking whisky or pastis. Another factor in the decline of Vin Doux Naturel may have been Portugal's entry into the European Union in 1986, which made for easier access to Port, often served as an aperitif in France. I remember early visits to France with my French exchange in the 1960s and her parents would offer their male guests whisky and the ladies a small glass of Port, not, I have to say, of the best quality. With sugar and alcohol no longer desirable, sadly sweet wines are not fashionable or popular any more. Vin Doux Naturel has acquired an image of a sweet wine for granny.

With the decline in the popularity of Vin Doux Naturel, prices plummeted. According to Georges Puig, in 1955 one hectare of Vin Doux Naturel produced 45 hectolitres, which was sold for a month's salary, of 800 French francs. Today that labourer would earn the basic wage, the SMIC, of €1,200, but a hectolitre of Vin Doux Naturel costs just €200. Between 1980 and 1990 Georges' father produced 30 hectolitres yearly from a hectare of Vin Doux Naturel, which equalled a week's salary. Today Georges obtains 17 hectolitres from a hectare, which is worth a day's salary, €100. Another figure I was given was one hectolitre of Vin Doux Naturel selling in 1982 *en vrac* for 1,560 francs, equivalent to €240. Georges is very pessimistic about the state of viticulture in Roussillon, considering it to be in a disastrous condition. It is badly affected by competition from Spain and Portugal, and the Greek island of Samos, with its reputation for Muscat. A wine estate was once worth a fortune, but here that is no longer the case. Emmanuel Cazes observed that at one time his father, Bernard, was single-handedly selling more Vin Doux Naturel than the entire AdVini group today. However, he is convinced that there is a place for the *haut de gamme*, the top of the range, such as their Cuvée Aimé Cazes Hors d'Age, and there is no doubt that Vin Doux Naturel has its aficionados among both the wine growers and the consumers.

Philippe Bourrier believes that Vin Doux Naturel should have a vintage on the label, which would help with sales. At the moment, an Hors d'Age could be anything between six and forty years old, sometimes even older. For him the vintage would provide a guarantee of age, and with it a guarantee of quality. There is still too much cheap Vin Doux Naturel, a legacy of when so much of it was sold to Byrrh in bulk. To develop an identity, Philippe believes, Vin Doux Naturel must express the grape varieties and the terroir. People also need to take more pride in their Vin Doux Naturel: 'If you are going to sell your wine, you must be proud of it,' he says. I recall an experience from an early visit to Roussillon in the late 1980s. I had spent a very enjoyable hour or so with an elderly wine grower, tasting his old Rivesaltes. He then suggested an aperitif before I left, in his house, rather than in the cellar.

Imagine my shock when he offered me a choice of whisky or Martini – 'but aren't I allowed some of your old Rivesaltes?' was my reply. He was thrilled, but it had simply not occurred to him to suggest his own wine. That is not an uncommon situation even today. Apparently, the cooperative members tend to drink pastis or whisky at meetings and the *apéro* of choice for the local rugby players is also usually whisky or Ricard, not Vin Doux Naturel. The local mayors serve beer, whisky or pastis for the *apéro de la mairie*.

There are as many opinions as there are wine growers about the future success of Vin Doux Naturel. For Pierre Boudau, Vin Doux Naturel was once a trump card, but became a weakness as people began to take it for granted. Instead of nurturing it – he used the French word *chuchoter* – it has become hackneyed or *galvaudé*, thanks to a sad short-term approach. Wine was once considered noble in Roussillon for it brought money, but even then people were not proud of it. The family vineyards were kept for the less bright child, who became a member of the cooperative. For Pierre, there is no reason why there cannot be a market for Vin Doux Naturel. You must give it back its *lettres de noblesse*, its credentials, he states. It must not be too cheap, so that people realize that they are drinking a treasure. Allowing the cheaper wines to disappear and nurturing the best would allow them to be *mis en valeur*, that is, properly appreciated and understood. Bruno Ribière has stopped making Vin Doux Naturel altogether, observing that while the wines are fabulous, *extra*, they are too complicated commercially. Vin Doux Naturel depends on élevage, and yet these wonderful old wines are sold relatively cheaply. Bruno prefers to concentrate on vin sec.

Price is indeed a complicated factor. If Vin Doux Naturel is too expensive it will not sell in the supermarkets. Domaine Augustin offers a 17-year-old Rivesaltes for €14 as the core of its range of Vin Doux Naturel. Hervé Lasserre, the previous sales director of Vignobles de Constance et du Terrassous, simply said, 'I don't know how to sell them any more expensively.' He called it 'un travail de fourmi', the work of an ant, but has succeeded in selling to the US and a little to Asia. These days Vin Doux Naturel sells better abroad than in France. With the availability of older vintages of Rivesaltes and also Banyuls, there are some fabulous wines at very modest prices. They simply need to be better known. It is important to reach out to the new generation. Jean-Pierre Papy, the director of the Rivesaltes cooperative is adamant: 'We have to fight to get these wines known; they must not be forgotten.'

Confusion over what type of drink Vin Doux Naturel is also makes it difficult to sell – is it an aperitif or a dessert wine? With the current trend towards drier aperitifs, white wine or even rosé, Vin Doux Naturel is rarely drunk in a restaurant. Julien Jeannin suggested offering *un verre gourmand*, rather than a *café gourmand* at the end of a meal, with little *mignardises*, or sweets. Vin Doux Naturel for Simon Dauré is a fantastic heritage, but it is a niche market and difficult to sell, and does not correspond to the taste of the current market. Daniel Laffite of Domaine des Soulanes laughingly observed: 'The thing I am most proud of is to have sold Vin Doux Naturel to a wine shop in Portugal!' Another problem is that Vin Doux Naturel may not be considered healthy, as it is relatively sweet and high in calories. It is often positioned on the market as a wine for celebrations, with an emphasis on either older vintages, or Muscat de Noël, the fresh Muscat of the year. For Marie-Pierre Piquemal, you have to really fight to sell Vins Doux Naturels, and you need the appropriate food to accompany them. Jean-François Rière thinks that many of the wine growers do not know what treasures they have in their cellars. However, he did admit that he barely drank one bottle of Muscat de Rivesaltes during the year.

On a slightly more positive note Jean-Marc Lafage considers that progress is being made with Maury Ambré. Even so, despite being one of the largest producers of Roussillon, his Vin Doux Naturel market is tiny, equalling just five per cent of his production.

Development of Rimage and Grenat

Rimage for Banyuls and Grenat for Maury and Rivesaltes are relatively new concepts, entailing young, fresh red wine, intended for early drinking. Banyuls is becoming increasingly associated with Rimage, which is more immediate, while in contrast the oxidative wines are more *recherché*. Grenat for Rivesaltes is the less important of the two. Bruno Cazes considers that Rimage can save Banyuls and also Collioure, and that old Banyuls is becoming rarer, even taking into account the fabulous stocks of old wine he rescued at the cooperative l'Etoile. White Banyuls is a minuscule part of the market, and pink Vin Doux Naturel remains in the doldrums.

Muscat de Rivesaltes

Muscat de Rivesaltes is complicated. The wines have not been appreciated for a long time, and so the producers have tried to diversify, creating a dry wine, a Muscat Sec as an alternative to Chardonnay or Sauvignon Blanc. Sparkling wine is another option being explored. However, Muscat is hard to sell; the flavours can be too overwhelming. There is no doubt that sales of both Muscat de Rivesaltes and Muscat Sec are falling, with a sharp decline in the vineyards of Muscat de Rivesaltes, from 4,756 hectares in 2010 down to 3,033 hectares in 2019. Pierre Boudau has put his Muscat de Rivesaltes in a smart bottle to make it look more of a premium product and increased the price accordingly, pointing out that one can only do this if the quality is there in the first place, and even then it still does not sell large quantities. Jean-Marc Lafage said that Muscat de Rivesaltes has been in freefall for some time. It used to be the *equilibre financier,* providing the financial balance of most estates, but this is no longer the case.

Rancio Sec

Rancio Sec was once an important part of the wine culture of the region; I was told that Collioure was originally a Rancio Sec, an unfortified wine aged in barrel without any *ouillage*. But as an appellation was never requested for it in the 1970s, it was in serious danger of disappearing altogether. Happily, a revival of interest led to its inclusion in the newer IGPs, Côtes Catalanes and, in 2012, Côte Vermeille. These days it is enjoying a revival of fortune, with a growing interest that needs to be developed. When asked about future projects more than one wine grower stated an intention to start producing a Rancio Sec. In recent years there has been a Salon du Rancio Sec; the fourth Salon Be Ranci took place in 2019 and included not only dry oxidized wines from Roussillon, but Rancio Sec from as many as 20 different regions from seven European countries, including Spain and Italy, as well as the French regions of Jura, Jurançon and the Loire Valley.

ROLE OF THE COOPERATIVES

The village cooperative was once a vital part of the village economy and the local community, but over the years the role of cooperatives has diminished. The number of cooperatives is falling pretty dramatically. In 2010 there were 50 cooperatives in Roussillon; today there are 24. An observation was made that there are two speeds of viticulture in the region: the cooperatives and independent wine growers. While that may be true in some cases, in my experience it is a rather unfair generalization. The cooperatives I visited, and obviously I was guided towards the more quality-conscious organizations, were pretty dynamic, although they are inevitably constrained by their more conservative members. While the good cooperatives are improving and moving away from the notion of volume, they still pay by the kilo and the alcoholic degree, although there are bonuses for better quality fruit or a better run vineyard. The average age of a cooperative member is approaching retirement, presenting the problem of who will take over the vineyards as members retire. However, the retirement of older members has opened the way to newcomers, allowing them to buy vineyards, often old vineyards, at an affordable price and set up their own estates. I was given an average figure of €8,000 for a hectare of vines, while old vines might be even cheaper, at €5,000. One wine grower paid €25,000 for two hectares of Grenache in good condition in 2018. Some of the independent wine growers see the cooperatives as complicating the situation when it comes to marketing the region's wines. Volume production coupled with the need for a fast turnover (they often have financial problems) means there is rarely a philosophy of quality.

ECONOMICS AND YIELDS

Since 1989 10,000 hectares of vineyards have disappeared from Roussillon, so that Roussillon now totals under 20,000 hectares. In 1989 1,800,000 hectolitres were produced, but by 2019 the figure had fallen to 622,321 hectolitres. Olivier Saperas observed that the small estates are no longer viable; they need to work together or disappear. Production costs are too high for small producers, and they are producing less because of the prevailing drought conditions. The production overheads in low-yielding years, when less wine is made, are essentially the same as in years with better harvests. There are about 15 estates in Banyuls and Collioure that are all in the same boat. An estate needs to be between 15 and 20 hectares to be viable – Olivier's grandfather had yields averaging 40 hectolitres per hectare; he, in contrast, obtains 15–20 hectolitres per hectare. You can't earn your living with such low yields, says Olivier, who also pointed out that while the Port producers have known for many years how to add extra value to their wine, 'we do not know how to do that. The Port producers are our greatest enemies.' He will not encourage his children to become winemakers.

Jean Boucabeille, who concentrates on vin sec, also observed that everyone is short of stock and that businesses are no longer viable with such small crops. A yield of 40 hectolitres per hectare is good and from that you can earn a decent living, but more often than not the average yield is between 10 and 25 hectolitres per hectare. Bulk prices are dropping, especially for Vin Doux Naturel, with a figure of €160 per hectolitre for three-year-old Ambré and €250 per hectolitre for Maury Sec.

There are vineyards for sale, not necessarily conspicuously, but every vineyard has its price, as Hervé Bizeul from Clos des Fées astutely observed.

EVOLUTION

There are changes for the better. Roussillon is gaining recognition as a distinct region, stepping out from the shadow of the Languedoc and becoming better known in its own right. You will quite often find the mention of Catalonia du Nord or Catalunya Nord on a wine label, emphasizing that Roussillon has more in common with the winegrowers to the south of the Pyrenees than those of Occitanie.

One of the biggest changes is the growth in the number of independent wine growers, with a greater variety of wines and a reorientation towards vin sec. Outsiders approaching Roussillon with fresh eyes are ready to question accepted practices, and are creating a much-needed dynamism in the region. The new generation of local wine growers is also playing its part in challenging the status quo – there are young people with new ideas. Gérard Gauby said there had been an explosion in the number of new wine cellars, especially in Maury, Calce and Latour-de-France. The cooperatives, in contrast, have slowed down. For Gérard, newcomers like Domaine le Soula and Domaine Thunevin-Calvet have demonstrated that the region can produce great wines and create opportunities for other producers.

For Eric Aracil, responsible for the export department of the CIVR, a role that includes communications with foreign wine writers, Roussillon has mostly shed its image of volume; wine growers are working on the quality of their wines, with a keen awareness of the different terroirs and of the quality of the hillside vineyards. That was certainly borne out in conversations with the various wine growers included in this book.

In 1980 the region comprised 50,000 hectares of vines, which produced 2 million hectolitres of wine; in 2019 19,674 hectares produced 622,321 hectolitres. When the vineyard area declined it was not necessarily the poorer quality vines that disappeared, that is, the higher yielding vineyards on flatter land, but often the older, less productive vines, which were the most difficult to cultivate. A wave of subsidies for *arrachage*, to pull up vines, meant that many vineyards on the edges of villages were replaced by new houses. Xavier Ponset at the cooperative of Baixas talked about the aftermath of the crisis in Vin Doux Naturel in the 1990s, when the cooperative joined Vignerons Catalans. The volume of Roussillon wine production dropped with the grubbing up of vineyards, but the number of wine estates increased. The area has become more dynamic; vineyards are cheap, attracting outsiders and there is a snowball effect. He also mentioned the growth in organic viticulture and the development of *vin nature* as positive changes in the region. For him Vin Doux Naturel is *une richesse, une identité*, an asset that is part of the region's identity.

There are still older vines to be found, and Cédric Haague at Vignerons Catalans enthused about Roussillon's new-found appreciation of these vines and the general effort being made to improve quality in the vineyard and cellar, with work in particular on the traditional grape varieties of the region. The market has evolved for the better but yields are still a problem. With four years of small crops, there is a shortage of grapes.

Changing the approach and outlook of producers across the region can be difficult. David Costa, the previous winemaker at Château Lauriga, took a dismal view of the prevailing attitudes in Roussillon, observing that it is very hard to get people here to change. They often lack an *ouverture d'esprit*, an openness of mind, and are led by jealousies; the mentality, the attitude of people can be a brake, he says. For David the department has everything going for it: the sea, the mountains, the grape varieties, if only attitudes could shift. 'We need to cut the chains, so that people can fly, *se débarrasser des mentalités*,' he says. Victor Gardiés noted that the independent personality of those in the region means that there is no collective spirit. As an outsider, Philippe Gard of Coume del Mas is a dispassionate observer who feels that people are not transparent. He believes tensions between growers mean that they do not work together, and they lack a long-term vision.

THE ROLE OF THE APPELLATIONS

Mickaël Sire is very committed to the region, feeling strongly that he cannot produce his best wines as IGP or Vin de France: 'I am from Roussillon, I am proud of my region. Even if Vin de France allows for more creativity and flexibility, we must *revendiquer*, or claim the region, the territory of Roussillon and its appellations.' However, he is critical of the endless meetings of the various committees; political machinations as he sees it. According to Mickaël the young generation, of which he is a member, is not interested in all that, even though they are the ones creating the new wave of wines. He considers that it is wine growers who will create the reputation of the region. He believes Roussillon has potential to achieve much more. Things have at least improved since his father ran the business, when *cavistes* would not even bother to taste the wines of the region, such was its lacklustre image.

While Olivier Pithon enthused about the variety of *cépages* and terroirs, he does not think that the appellation Côtes du Roussillon means much to wine consumers anymore, and at the same time wine growers would prefer the freedom to create the wines they want, without being tied to the regional appellation. Similarly, all Jean-Philippe Padié's wines are Vin de France. The white wines were originally IGP Côtes Catalanes, but as he put it, he did not want to be Don Quixote and 'fight the windmill'. He did not recognize himself in the appellations and has not belonged to the CIVR for years. While he acknowledges that it can be a handicap not to be a member in terms of communication with other growers, he does have freedom. In any case, he notes, the image of the appellations is not brilliant; they are more for big companies and supermarkets. The producers in Roussillon seem to fall into two broad camps: those that conform and support the appellations and those who will have nothing at all to do with them, such as the natural wine producers. Orange wines do not fit into the appellations either and are either IGP or Vin de France. Jean-Philippe considers that Roussillon is just waking up, thanks above all to the white wines of the region, and to a slow but growing interest in *rancio*. For him the key assets of Roussillon are the heritage of old vines, the terroir and the salinity in the wines.

Gérard Gauby gave up on the appellations when he was refused a derogation for a wine that was too low in alcohol, with the suggestion that he remedy the 'problem' by adding some concentrated must. I can imagine the expletives by way of reply! As a consequence, all his wines are now Côtes Catalanes. In contrast Eric Laguerre, of Domaine Eric Laguerre, remains loyal to the appellation system, commenting that his father and grandfather fought for the appellation, as they considered that with it the wines would sell better, which of course is not always the case today. However, Eric respects their work and continues to follow their example.

Although Côtes Catalanes is the principal IGP of Roussillon, some people, such as Simon Dauré at Château de Jau, prefer IGP Pays d'Oc, which Simon uses for Chardonnay, observing that people do not know Côtes Catalanes. He would like to see more aromatic varieties in Côtes du Roussillon, maybe even including Chardonnay. If your wine is not an appellation wine, you cannot use château on the label, even if you have a bona fide château, but he does not think that an appellation helps sales. Simon has always enjoyed marketing, and likes taking risks while being innovative. 'Our metier is to make wine more accessible,' he observed. 'Roussillon is a land of milk and honey for wine.'

In the current (fourth) edition of *The Oxford Companion to Wine*, edited by Jancis Robinson MW, a quote about Roussillon caught my eye: 'The vine-growers of Roussillon have been some of France's least content with the details of their appellation regulations, which continue to evolve. There can be considerable scepticism about a system devised as far away as Paris and administered from Brussels, especially among those who identify so closely with the inhabitants of Barcelona.' This idea was borne out by my conversations with the region's growers.

So why is Roussillon so little known? Why does it not have the cachet of Priorat? Like those of Roussillon, the fortified wines of Tarragona had a reputation, which has been superseded by the neighbouring table wines of Priorat. Why has Roussillon so far been unable to make a similar shift in reputation? No one has really been able to answer that question. One suggestion was the influence on Priorat of nearby Barcelona, an international city, as opposed to Roussillon's major town, Perpignan, which is a provincial prefecture. Taking into account the views of the growers noted above, maybe the individualism of the wine growers has a part to play. As yet Roussillon has not had any market leaders on the international stage. Gérard Gauby and Mas Amiel have been important pioneers in the region, but for some reason they do not have the same international recognition as somebody like René Barbier at Clos Mogador, and others who followed his lead.

WINE TOURISM

Perhaps the development of wine tourism will help the reputation of Roussillon. One regional wine event is the Amorioles de Maury, which usually takes place on the third Sunday in May. It is what is termed a *balade gourmande* around the vineyards of Maury, a walk of about 7 kilometres with regular refreshment points for wine and food. When I took part I enjoyed it enormously, as an opportunity to taste a broad range of wines and to meet some new wine growers. Noël Lafforgue at Domaine Lafforgue is planning some *balades equestres* as his partner is a keen and talented horsewoman and they rightly feel that this would be a wonderful way to introduce people to the region. He promised me a ride with a very calm pony, when I next return.

Of course, wine goes with food, and there are several wine estates with restaurants. I can personally recommend Les 9 Caves in Banyuls, and Riberach in Bélesta, as well as Domaine de Nidolères outside Tresserre. The menu at Clos des Paulilles looks very tempting and there is also a restaurant at Château de Jau, all of which show off the region's wines and gastronomy. In addition, visits to cellars without restaurants can easily be arranged with a phone call or email. Provided there is nothing pressing in the cellar or vineyard, most wine growers are very happy to talk and taste, and hopefully sell.

Tourism should be helpful in addressing one of Roussillon's main challenges, the fact that it is still relatively unknown as a distinct region. Its heterogeneity, with so many different terroirs, estates and styles of wines, makes it difficult to apply a universal identity to the region and means that newcomers, discovering the region for the first time, may find it confusing. However, its variety is an important part of its charm. You can find almost every flavour of wine that you might imagine within one relatively small area and the quality of the wines has improved enormously in recent years. Eric Aracil of the CIVR confidently sees the future as rosy, and as Jean-Marc Lafage put it, travel makes us appreciate what we have here. Richard Bray of Cathar(tic) Wines commented on the amazing raw material in the region and is convinced that the ceiling of quality has not yet been reached.

I am very aware that some of the picture I have painted in this chapter is far from rosy. Roussillon does have problems, but it also has a growing core of wine growers who are deeply committed to their region, fully appreciating its assets and prepared to fight for them. The vins secs will undoubtedly grow in reputation as they gain wider recognition, and hopefully there will be also be a revival in the reputation of the finer Vins Doux Naturels. I do hope so; the quality is undeniably there, and the aged *rancio* wines stand amongst the vinous treasures of France.

PART 2

WHO'S WHO IN ROUSSILLON

3 Producers in Part 2 are organized alphabetically within their respective communes.

6

MAURY, ST PAUL AND THE NORTHERN END OF THE AGLY VALLEY

MAURY

Most wine estates produce more than one appellation and so there is considerable overlap, making the organization of a directory of producers somewhat challenging. I have decided to start in the village of Maury, as that is where my research began, and then gradually fan out along the Agly Valley, with other clusters of villages and then head south through Les Aspres to reach Banyuls and Collioure.

Maury is a long, thin village in the centre of the Agly Valley. As you approach it from the east, you see the outline of the ruined Cathar castle of Quéribus and the old lookout tower of Tautavel. At one end of the village is the modern face of the cooperative, with a smart shop and a charming mural of three elderly wine growers sitting on a bench. Opposite, there is the friendly restaurant Le Pichenouille, where you are more than likely to encounter a vigneron or two. At the far end of the village there are more large buildings also belonging to the village cooperative, cellars, with some enormous stainless steel vats for vinification, and a warehouse for the storage of bottles. The adjoining train station was created specifically for the delivery of grape juice, with a train running from Rivesaltes. These days it is a tourist service, *le train rouge*, running as far as Axat, at the end of the valley. The name is a bit of a joke: the formal name of the service is *le train du pays Cathar et des Fenouillèdes*, which shortens to *le train du PCF*, which more commonly in France stands for the Parti Communiste Français. Turn off the main street and you will find narrow streets leading up to a small square and a modern church. And every now and then you will see an eye-catching mural painted by local artist Bernard Gout. I stayed in a friendly *chambres d'hôte* at the top of the village, with wonderful views from the terrace over the rooftops of Maury to the hills behind.

The appellations of Maury and Maury Sec form a small enclave within the larger appellations of Côtes du Roussillon Villages and Rivesaltes. The Vin Doux Naturel, Maury, was recognized as an appellation in 1936 but the appellation for the table wine, Maury Sec, has only existed since 2011. The twin appellations are based on schist, and also cover the neighbouring villages of Saint-Paul-de-Fenouillet, Tautavel and Rasiguères as well as Maury itself. Where the schist ends, the hills of the Corbières begin and only a little further east down the valley, the next village, Estagel, is part of Côtes du Roussillon Villages. Maury itself is quite diverse with a marked difference between the two sides of the valley. The south is hotter, so the vineyards tend to face north and there is more influence from the Mediterranean and the Pyrenees. On the north side of the Agly there are more valleys, with more undulating countryside.

Mas Amiel

Maury

www.masamiel.fr

The leading estate, which was at one time the only serious estate in Maury, is Mas Amiel. The estate has had a colourful history. The story goes that a certain Raymond Amiel won the title deeds for the property from the Bishop of Perpignan at a game of cards back in 1816, when the estate amounted to some 80 hectares, mainly of *garrigue*-covered hillsides and some vineyards. At that time, the property was largely devoted to rearing rabbits, which were sold in the Perpignan market. The vineyards were replanted with grafted vines in 1894 after the phylloxera crisis, by the son of the card player and an associate, a négociant, named Camille Gouzy. At the same time, they built the cellar. M. Gouzy was bankrupted by the viticultural crisis of 1907 and died, ruined, in 1910. M. Amiel was forced to mortgage the estate, which was then bought by Charles Dupuy. His son, Jean, did much to develop the wines, breaking a contract with Byrrh in order to produce his own Vin Doux Naturel, which he first bottled in 1924, in bottles displaying a stylish Art Deco label.

I first visited Mas Amiel in 1987 and met the then owner Charles Dupuy, son of Jean Dupuy. When Charles died in 1997 Mas Amiel was bought by Olivier Decelle, who was running the big frozen food company Picard at the time. He has since sold that to buy estates in Bordeaux, Châteaux Jean Faure in St Emilion, Haut Maurac in the Médoc and Haut Balet in Fronsac, as well as six hectares of Savigny-lès-Beaune and Marsannay. He also has a négociant business dealing in about 20 Burgundy *crus*. Since 2003 the wine at Mas Amiel has been made by Nicolas Raffy, who initially worked as a biochemist, but changed direction to study oenology at Montpellier, gaining experience at Château Montus, the leading estate in Madiran, before coming to Mas Amiel.

Until 1997, the estate produced only Vin Doux Naturel, but Olivier Decelle instigated the production of vin sec, with Vin Doux Naturel now accounting for just a quarter of the production. One of the most dramatic things you see as you arrive at Mas Amiel is a large *parc de bonbonnes*, the large glass jars used for ageing the oxidative Vin Doux Naturel for 12 months. Even if production has declined, there are still 800 *bonbonnes*, exposed to all the elements; the heat of summer, the wind and the rain. There is also a wonderful cellar of old foudres built in 1890 – both the cellar, and the barrels in situ – 17 magnificent 250-hectolitre beasts. They are made of chestnut rather than oak; apparently spiders do not like chestnut and there are indeed no spider webs in that cellar. Otherwise the cellars at Mas Amiel have been well modernized. There is an enormous variety of different containers: concrete vats, including small horizontal ones, barriques, tronconic oak vats, amphorae and some small free-standing concrete vats that they call the sputniks, after their appearance.

Altogether there are 145 hectares of vines, in 133 different plots, all around the property, on the slopes of the Agly Valley, and towards Tautavel, and with one small plot at Montner. All the vines are farmed organically, and are in conversion to biodynamic viticulture. They make 25 different wines, some blends and some from specific plots. The soil is mainly schist, but there is also some clay, limestone and granite. The red grape varieties are Grenache Noir, Syrah and Carignan, while for white wine there is Grenache Blanc, Grenache Gris and Macabeo, plus more recent additions Marsanne and Roussanne, and Muscat for Muscat de Rivesaltes.

Winemaker Nicolas Raffy was away on my most recent visit to the estate so I tasted with Lucas Gabor, who works in the visitor centre. He comes from Alsace but has fallen in love with Roussillon, and is well-informed on recent developments at the property. I was treated to a comprehensive tasting, giving me a very good overview of the wines of Mas Amiel. Highlights included a Côtes du Roussillon Blanc, a blend of Grenache Blanc, Macabeo, Roussanne and Marsanne kept in vat, with some herbal notes. In contrast Altaïr is a field blend of 40-year-old Grenache Gris, Grenache Blanc and Macabeo from limestone soil, kept in demi-muids for less than six months. The grape varieties can be vinified separately as they are quite distinguishable in the vineyard.

Vertigo Côtes du Roussillon, a blend of Grenache, Carignan and Syrah, is ripe and spicy with supple tannins, while Initial Maury Sec, although a similar blend, had some weight and depth. Origine is a Côtes du Roussillon Villages from one-third each of Grenache, Carignan and Syrah, given two years in wood. Terres Rares is a Maury Sec, from one small plot of Grenache and Carignan, combining the freshness of Carignan and the cherry fruit of Grenache. Alt.433M is from the vineyard at Montner, set at 433 metres altitude, on mica schist and granite, and planted with Grenache and Lledoner Pelut. With an alcohol level of only 12.5% abv, which is relatively rare for Roussillon, it was fresh and fragrant. Vers le Nord comes from vines on the hills below the Cathar castle of Quéribus, and consists mainly of Grenache with a little Syrah, with no oak ageing. Voyage en Météore is a Maury Sec, a blend of Grenache and Carignan from the most northern vineyard, which receives the least sunshine. It had appealing fresh fruit.

A delicious assortment of Vins Doux Naturels followed. Vintage Blanc, from Grenache Gris, is soft and floral; Muscat de Rivesaltes is a blend of 80 per cent Muscat à Petits Grains and 20 per cent Muscat d'Alexandrie, with some restrained fruit. The 2015 Vintage Maury, a pure Grenache Noir, *muté sur grains*, was rich and spicy. A 2012 Charles Dupuy, from 100-year-old Grenache vines, had spent 18 months in the big foudres, so that it had a rich, chocolatey nose, and on the palate was ripe, but not sweet, with great depth.

The real stars were the oxidized wines. A 30-year-old Maury, which, rather than indicating an average age, entails a minimum of 30 years of ageing in *bonbonnes* and then foudres, was elegantly nutty. The 40-year-old wine was more mineral with less sucrosity, but very intense, with good acidity. Next came Les Vins d'Héritage, a selection of older vintages. I tried the 1980, a year of good acidity with lovely concentration, elegance and length, and then the 1969 which had been bottled in early

2019. It was elegantly nutty, less intense and more subtle than the 1980, making a delicious finale to a very impressive tasting. There is no doubt that Mas Amiel is an excellent ambassador for Maury and for the whole of Roussillon.

Domaine of the Bee

Maury

www.domaineofthebee.com

I have known Justin Howard-Sneyd MW through the British wine trade for a number of years. During lockdown in 2020 we met via Zoom to discuss how he came to develop Domaine of the Bee. He had realized fairly early on in his wine trade career that he would eventually like to make his own wine. Buying vineyards outside Maury was the result of a series of circumstances; the loan of a house in the Corbières and the opportunity to discover the Agly Valley over the other side of the hills, and then a phone call to say that a vineyard was for sale. So, in February 2004, Justin bought a 1.8-hectare block in the hills on the south side of the Agly Valley, just outside Maury, close to the vineyards of Clos del Rey. The 100-year-old vines are *complanté*, mainly Grenache in all three colours, with Carignan, both red and white, and the odd Macabeo and Muscat vine. The original name was intended to be Domaine de l'Abeille as the Roc de l'Abeille is nearby. However, there was an objection from a Monsieur Abeille in the Rhône Valley, and so the name was anglicized, with accompanying humour for the various cuvée names, The Bee-Side, Bee Pink, Les Genoux ('knees' in French) and so on.

Initially Richard Case of Domaine de Pertuisane looked after the vineyard and the wine was made in his cellar, but once Richard teamed up with Dave Phinney, who did not, as Justin puts it, 'want the refugees', they moved to Jean-Marc Lafage's cellar at Château St Roch between Maury and Estagel. Justin now has four hectares, having bought two more blocks, on a north-facing slope outside Maury, and he also buys grapes for white and rosé from Jean-Marc. Although Justin lives in London, he is always in Maury for five weeks over the harvest.

His winemaking methods are quite straightforward. The wine is fermented in 500-litre oak barrels, with a pre-fermentation maceration. He usually keeps some stems as they add freshness and lower the alcohol level. Each individual plot is separate, allowing for blending opportunities. Initially he made just one wine, but the range has gradually evolved. Field of the Bee, a white Côtes Catalanes, comes from Grenache

Blanc and Grenache Gris; Bee Pink is a pink Côtes du Roussillon from Grenache Noir, Grenache Gris and Syrah. The Bee-Side is a lighter red wine, based on Grenache, with rounded fruit. The pure Carignan, Côtes Catalanes, from 80-year-old vines, is a lovely expression of that variety. Domaine of the Bee, a Côtes du Roussillon Villages blend of Grenache and Carignan, is the principal wine of the estate. Les Genoux comes from the oldest block of vines, La Coume du Roy, and is only made in a good Grenache year, such as 2018. It is a lovely example of Grenache at its most elegant.

Justin has also experimented with orange wine, making Taronja (orange in Catalan) with Jean-Marc Lafage. It comes mainly from Grenache Gris and Grenache Blanc, with a little Muscat and even less Viognier, with four different barrel fermentations in 2018. The wine had a youthful freshness, balanced by some dry tannins, and pithy fruit with a refreshing finish.

Domaine des Enfants

Maury

www.domaine-des-enfants.com

Domaine des Enfants is a joint venture between Carrie Sumner and Marcel Bühler, with a new cellar on the edge of the village. Carrie is an American, from Portland, who worked as a sommelier in New York and made wine in the Napa Valley before coming to Roussillon to work on a project. Here she met Marcel, who comes from Switzerland and had been a banker in Zurich. Tiring of banking, Marcel went to culinary school in Paris, studied winemaking at Geisenheim and fell in love with Priorat; Roussillon was the next best thing. Despite being a mathematical genius Marcel really always wanted to be a gardener, so he now tends his vineyards and Carrie makes the wine. As Carrie put it, he has returned to his childhood dream, which accounts for both the name of the estate and the names of the various wines.

Marcel made his first wine in Maury in 2007, and Carrie's first vintage was 2010. They have 26 hectares altogether, in about 23 different plots, with different soils and microclimates, all along the Agly Valley. There is Grenache in Maury and St Paul-de-Fenouillet and Syrah in Rasiguères and Caramany; last year they planted some Mourvèdre, and they are also planting olive trees, and more white varieties. Our tasting began with Le Jouet Blanc, a pure Macabeo, with firm, saline fruit, tight-knit and textured with great length. They crush the grapes by foot, press them, protect them from oxygen with carbon dioxide and, after *débourbage*, ferment them in a stainless steel vat. There is no malolactic fermentation and just a little *bâtonnage* and the wine is bottled early, in November. Tabula Rasa ('clean sheet'), is a field blend of 30 per cent each of Macabeo, Grenache Gris and Grenache Blanc, with 10 per cent Carignan Blanc. The grapes are crushed and given a cold maceration overnight and the wine spends 11 months on the lees in 400-litre neutral barrels. The wine had dry honey on the palate, with weight and texture and a very satisfying finish.

Le Jouet Rosé is deep coloured, made mainly from Syrah, with Grenache and Carignan – they limit the amount of Grenache in consideration of its high alcohol level. The grapes are picked early and the wine is quite mouth filling, with dry, cherry fruit. Le Jouet Rouge is a blend of their younger vines, Grenache Noir and Lledoner Pelut with a little Carignan and Syrah, fermented partly by carbonic maceration and aged in vat for 12 months. It is a lighter red for earlier drinking. L'Enfant Perdu is a blend of equal parts Grenache, Carignan and Syrah, aged for 14 months, including 15 per cent in new oak. It was rich and spicy with fresh tannins. Sous l'Etoile, Vin de France de Catalunya Nord, a blend of 40 per cent each of Carignan and Grenache Noir, with some Syrah and a drop of Mourvèdre, was aged partly in old 400- and 600-litre barrels for 12 months, with the Syrah kept in smaller Burgundy barrels. This is wine to age, with a deep colour, firm fruit and structured tannin. La Larme de L'Ame comes mainly from Syrah with a little Carignan and Grenache. Carrie explained how they are experimenting with the stems by filling a stocking full of stems and dunking it rather like a teabag. The wine was closed and tight-knit, with firm fruit and well-balanced tannins.

Carrie is also working on wines of her own, studying pure varieties on a particular soil. She has made Carignan from schist, Grenache on granite at Bélesta and then she may blend them. 'I am not good at following the rules, so these wines are Vin de France,' she says. Chroma Soma was a blend from two different Grenache vineyards, one schist, one granite, plus some Syrah grown on granite, and Carignan from granite at Montalba. The flavour was smoky with some integrated oak. A Grenache from Caramany, after a cold soak, a whole bunch crush and an élevage in a sandstone egg, had some elegant tannins and round, silky fruit, and not a trace of wood. As Carrie said, this way she learns about the microclimate and the soil. I enjoyed her enthusiasm, while Marcel is more restrained in his approach.

Mas Janeil

Tautavel

www.masjaneil.com

Mas Janeil is the Roussillon estate of François Lurton, with a first vintage in 1999. When I visited I met Julien Fernandez, who is responsible for François' vineyards of Languedoc-Roussillon – he also owns Château des Erles in Fitou – as well as the chief winemaker of the company, Xavier-Luc Linglin. First, we looked at the vineyards. There are 34 hectares in the valley of Soulanes, which runs between Maury and Tautavel, along the foot of the Corbières hills, at about 200–300 metres altitude. The soil is schist, with some limestone. Grenache accounts for 60 per cent of the vineyards, with Syrah, Carignan and Mourvèdre making up the rest of the reds. For white wine there is Grenache Gris and Macabeo, and Grenache Blanc and Vermentino have also been planted. For the IGPs there are also plantings on lower, flatter land of more international varieties, namely Viognier, Chardonnay and Sauvignon. The estate concentrates on vin sec, and also produces some *vinaigre de Maury*, which has its own appellation. They insisted we see their highest vineyard, Le Pas de la Mule, nestling under the hills dominated by the Cathar castle of Quéribus. The drive to reach it was challenging and Julien showed considerable potential for the Paris–Dakar rally as he negotiated deeply rutted tracks. From the vineyard we were able to look down on the little hamlet that gives Mas Janeil its name.

The cellar, just up the track from Domaine des Soulanes, is efficient and well-equipped, with an *égrappoir*, a sorting table and a cold chamber, which they use for a pre-fermentation maceration, to extract colour and aroma, but not tannin. They use concrete vats for red wine, and stainless steel vats for the white. There are also some plastic polymer eggs, which they can place in the cold chamber, if necessary, as well as concrete eggs.

They have restored a house in the hamlet, where we tasted. First came Le Petit Pas Blanc, a blend of Macabeo and Grenache Gris, with some fresh, stony fruit. Mas Janeil Blanc is also a blend of Macabeo and Grenache Gris, plus a little Grenache Blanc, fermented in barrels, barriques and demi-muids. It was quite ripe and honeyed with oaky tannins.

Le Petit Pas Rouge, from Grenache, with some Syrah, Carignan and Mourvèdre is kept in vat. It was supple and rounded. Xavier-Luc explained that they are moving away from oak. Up to 2015 they followed the bordelais system of a *grand vin* and a second wine, but now they have identified specific plots, so Le Petit Pas sees no wood, and the *grand vin* is Mas Janeil Côtes du Roussillon Villages, from Syrah, Grenache, Carignan and Mourvèdre. They make two wines, one with sulphur and one without, both coming from the same vineyards.

For the wine without any added sulphur, they sometimes use indigenous yeast but on other occasions they use selected yeast. The fermentation is a form of carbonic maceration, with a cool pre-fermentation maceration and a warm post-fermentation maceration, with the various grape varieties, in layers, with and without their stems. The wine was perfumed, with a rounded palate and some balancing tannin. Xavier-Luc observed that yeasts behave differently when they are not fighting sulphur. They pick earlier, so that the wine has higher acidity, which gives some protection against oxidation. The wine without sulphur contains equal proportions of the four varieties, whereas the classic Mas Janeil includes more Grenache Noir, with some Syrah, Carignan and Mourvèdre, and also entails some *vinification intégrale*, making it more concentrated, with spicy cherries and good structure.

Le Pas de la Mule is not made every year. It comes mainly from Grenache, with Syrah and Mourvèdre. We tasted the 2015, which was made when 'we were still looking for very ripe wine'. It was rich and extracted, with black fruit. Xavier-Luc wants to move away from wood, and is looking for purity of fruit. In 2019, they used amphorae for the first time. We left with Julien and Xavier enthusing about the 2019 vintage, 'un bonheur'.

Mas de Lavail

Maury

www.masdelavail.com

Mas de Lavail is a large estate on the road between Maury and Estagel. Jean Batlle, who had previously been a member of the Maury cooperative, bought the estate some 20 years ago. Today his children, Sophie and Nicolas, run the property. Nicolas is the winemaker and Sophie gave me a friendly tasting, explaining that 75 per cent of production from the 75-hectare estate is now vin sec; in contrast the previous owner produced only Vin Doux Naturel, and sold only *en vrac*. The *mas* is 200 years old, and Lavail is the name of the *lieu-dit.*

In the vineyard, they practise sustainable viticulture, using no insecticides and favouring pheremonic controls; they are working towards HVE and maybe eventually an organic certification. The greater part of their estate is planted with Grenache Noir, Grenache Blanc and

Grenache Gris, but they also have Carignan Noir, Syrah and a little Mourvèdre. There is also some Muscat for Muscat de Rivesaltes and they are replanting some Carignan Blanc.

Our tasting began with Ballade Blanc, a Carignan Blanc from a two-hectare plot of 80-year-old vines, which was fresh and stony. Le Sud, a Côtes Catalanes, is a blend of Grenache Gris and Blanc, from 60-year-old vines, fermented and *élevé* in wood, with oaky notes balanced by good acidity. The Syrah and Carignan for Ballade Rosé are picked early in the morning, pressed and allowed a malolactic fermentation, to make a ripe, rounded wine with good balancing acidity. Ballade Rouge comes from 50-year-old Carignan with some ripe fruit and a fresh finish. There are two contrasting Côtes du Roussillon Villages, with varying percentages of Syrah, Grenache and Carignan. La Désiderade is kept in old 300-litre barrels for 12 months and the oak was well-integrated. Sophie described their style as 'sur le fruit' – they do not like too much oak.

Ego, a Maury Sec, is mainly Grenache Noir, aged in demi-muids for 12 months, with elegant, dry spice. The final red wine was Initiale, another Maury Sec, mainly Grenache with 10 per cent each of Mourvèdre, Syrah and Carignan, fermented in new wood, and given 14 months of élevage in 300-litre barrels. It comes from the best vineyards and the palate is firm, structured and elegant.

Maury Grenat, *muté sur grains*, with 12 months in barrel, was ripe and spicy with an oaky, tannic streak. Maury Blanc, from Grenache Blanc, with 30 per cent Grenache Gris, *muté sur jus*, and kept in wood for nine months, was rounded and biscuity with a hint of aniseed. Sophie observed that the colour of the Vin Doux Naturel determines the choice of dessert, or vice versa, so with Maury Blanc she would favour a *tarte tatin*. A 1999 Maury Ambré Hors d'Age commemorates the first vintage of the estate. They put aside some Grenache Blanc as well as some Grenache Noir, *muté sur grains*, and completely forgot about it for 20 years. Two-thirds had evaporated, but the remaining wine was elegantly nutty, with an intense nose and palate, with candied fruit and a dry finish. It was utterly delicious and extraordinary value (in 2019) at €69.00 per bottle.

Mas Mudigliza

Maury

mas.mudigliza@neuf.fr

Owner Dimitri Glipa explained that the name of his estate is made up of alternating pairs of letters from his name and his wife's, Muriel

Sanson, but they turned an s into a z to create Mudigliza. Dimitri hails from Libourne, near Bordeaux, and was inspired to come to Roussillon by Caroline Bonville of Mas Karolina (see page 112); their parents are friends and he liked her wine. In December 2005 he asked her to find him five or six hectares in Roussillon. They came to look at the end of January, a visit which coincided with 60 centimetres of snow, but three weeks later they looked again, in sunshine. They made their first wine that year. They now have 12 hectares, half at St Paul-de-Fenouillet and half in Maury on the north side of the valley, and their cellar, a large functional shed, is in Maury, on the edge of the village. Explaining the differences between St Paul and Maury, Dimitri said that St Paul, further up the valley, is cool with some schist and marl, whereas Maury is hotter. He has recently bought a vineyard at Tautavel and altogether has about 30 different plots, making a veritable patchwork of vineyards and enabling him to blend and play with the ripeness levels.

Caudalouis, Côtes Catalanes, comes from 70-year-old Grenache Gris and Macabeo from a vineyard in St Paul, where growing conditions are influenced by the Pyrenees. The Macabeo is kept in a concrete egg and the Grenache Gris is fermented in barrel, making for good structure, minerality and firm acidity. The 2014, enjoyed in London when it was five years old, showed wonderful ageing potential. For the wine to be Côtes du Roussillon, Dimitri would need Grenache Blanc, Marsanne or Roussanne, none of which he considers interesting. La Coume des Loups, his entry level red wine, is a blend of Carignan, Grenache and Syrah aged in vat, with some fresh spice. Carminé is deep red, and made from 70 per cent Grenache, from St Paul and Maury, all aged in vat, with Syrah from St Paul, of which a part is aged in barrel for ten months. The components are then blended and given further élevage in vat, maybe for a year. It was redolent of spicy tapenade on the nose, supple, but with a backbone. Symbiosis, Côtes du Roussillon Villages, is a blend of two parts Carignan to one part Syrah, aged in 300-litre barrels, making for a ripe, rounded nose and palate, with a firm finish. Future vintages will be Maury Sec, now that Dimitri has some Carignan within Maury. He also makes a pure Carignan, from a vineyard near Galamus, which was firm and structured.

Dimitri is very enthusiastic about Vin Doux Naturel, considering it to have a better *rapport qualité prix* than Port. His Maury Grenat is kept in wood, and muted after about seven days, with 85–90 grams of sugar per litre. It had rich, red fruit, with a touch of chocolate and a streak of tannin. Maury Blanc comes from Macabeo *muté sur jus*, at 100 grams of sugar per litre, so that it is ripe and honeyed. Ultimately Dimitri would like to make an Hors d'Age from Muscat, which might possibly resemble an old Liqueur Muscat from Australia.

Clot de l'Origine

Maury

www.clotdelorigine.com

Marc Bariot comes from Bandol, but the significant difference in the price of vines between Bandol and Roussillon – €200,000 in Bandol compared to Roussillon's €10,000 per hectare – brought him to Roussillon. Back in 2004 when Marc started out here vines in the Languedoc were more expensive than in Roussillon and Gérard Gauby was demonstrating just what a great white wine region Roussillon could be – it seemed like a good opportunity. As a newcomer to the region, Marc is nicely opinionated about the region, observing it with the eyes of an outsider.

Marc has vineyards in Calce, Maury and further up the Agly Valley at Caudiès. With Carignan in Calce and Syrah at Caudiès, Maury is at the centre. He began with two hectares in 2004 and now has 12 hectares, and is working biodynamically. Marc observed that 'for the "good" Catalan vigneron grass is dirty', and sees organic viticulture as 'a fashion that will pass'. He took us to see his vineyards towards St Paul-de-Fenouillet, where he showed us his old vines; some had grown too tall, so he has shortened them, enabling them to better withstand the wind. We looked at weeds: 'a good weed is like an iceberg, with lots of roots which help retain the soil and combat erosion,' he says. Marc talked of a duvet of Gramineae, or grasses. He is planting half a hectare by massal selection, all three colours of Grenache, with Macabeo and Carignan Noir. He would also like to try Assyrtiko, but there is no subsidy available for that variety, an important consideration when it costs €40,000–€50,000 to plant one hectare, including the cost of the land.

Back at his rather cramped cellar in the village, we tasted some vat samples and some bottles, all from the most recent (2019) vintage. At Caudiès, he has Sauvignon Blanc, Syrah and Merlot, and in the two hectares at Maury there is Grenache in all three colours. At Calce there is Carignan Noir, Grenache Gris, Grenache Blanc and Macabeo, with some Muscat d'Alexandrie at Montner. With vineyards so spread out, there is a difference of a month in the picking dates, starting on 15

August in Maury and finishing at Caudiès on 25 September. Marc did admit that it was a bit mad. Although red varieties account for 10 hectares, most of his wine is vinified *en blanc*, so that 90 per cent of his production is white. This certainly was an original take on winemaking, which I have never encountered before.

Explaining his range, Marc told me that Le P'tit Barriot comes from the youngest vines. Next come Les Quilles Libres and L'Original and then Le Trouble Fait. But that was not the order of our tasting. It began with a Sauvignon, with some fresh, stony fruit. Next came L'Original, a white blend of 80 per cent Macabeo with 20 per cent Merlot. The Macabeo, from 60-year-old vines, retains acidity, and more comes from the Merlot, so there was indeed firm acidity, and some herbal iodine notes. Les Quilles Libres Blanc – *quilles* is a term for good bottles, and the name is also a play on words with *équilibre* meaning balance – comes from 80-year-old vines, from all three colours of Grenache, all three colours of Carignan, Macabeo and some Muscat. The Grenache Noir gives some *moelleux*, or sweet richness, the Grenache Blanc aroma and the Grenache Gris acidity. Everything is pressed immediately without any skin contact. A blend of equal parts Muscat and Syrah is used for an orange wine, Le Trouble Fait. Whole bunches spend a month in vat, with a daily *pigeage* and no temperature control. It was very intriguing, and Marc explained that the Muscat, fermented dry, gives complexity while Syrah adds acidity and tannin. 'And you can drink it with everything except French food,' he joked.

The first red, Le P'tit Barriot, comes from young Syrah aged for 12 months in foudre and then vat, giving fresh, peppery fruit. Soif de Plaisir is a blend of 80-year-old Carignan from Calce, planted on a steep, north-facing slope with some much younger Syrah vines, that has fresh red fruit and quite firm tannins. Carignan provides complexity and an identity, and on north-facing slopes you get ripe grapes without too much alcohol. Les Quilles Libres Rouge is a blend of 60 per cent Carignan with some Grenache and a little Syrah, with some elegant perfumed fruit. It spends nine months in a 500-litre Stockinger barrel.

Finally we moved on to Vin Doux Naturel. Maury Ambré, a blend of Grenache Blanc, Grenache Gris and Macabeo aged outside for three years in barrel, had elegant concentration, fresh nutty notes, and hints of bitter oranges. It reminded me of old Sauternes. A Rancio Sec from the same vineyards came straight from the barrel. It had also spent three years outside and was firm and austere.

Domaine de la Pertuisane and D66

Maury

www.pertuisane.com; www.department66.com

Richard Case studied wine at Plumpton College outside Lewes, England, at the end of the last century. Following his studies he went to Domaine de la Baume, near Pézenas, moving on to work in Gisborne, New Zealand, and then Chile, before buying his first vineyards in the Agly Valley at the end of 2002. He described his first vintage in 2003, as a baptism of fire, with both heat and hail.

Richard is also the general manager for an American venture, D66, named after the departmental number of the Pyrénées-Orientales. Richard met Dave Phinney through his US importer in 2008, and in 2009 made Dave's first wines from Roussillon. Dave was already an established winemaker in the Napa Valley, best known for The Prisoner, a label that he has subsequently sold to Constellation. A spacious state-of-the-art cellar designed by an architect from Bordeaux has been constructed, with numerous small stainless steel vats and two extensive barrel rooms where the barrels stand six high. Richard uses a small space for his own wines. He has 10 hectares, while Dave's 125 hectares make him one of the larger estates of Roussillon

We tasted the wines from both estates. A pertuisane, a Cathar weapon, the point of the spear, features on Richard's labels. The first of his wines, The Guardian, a Côtes Catalanes Grenache Gris, was fermented in stainless steel and then put in what Richard called neutral oak, barrels that are at least three years old. He observed that people used to pull up Grenache Gris, and now they are saving it, and even planting it. It is a field blend with a little Grenache Noir and there is a month's difference in the ripening, early September for Grenache Gris and the end of the month for Grenache Noir. The wine has good tension, with some rich fruit, after seven months in oak. D66's rosé, Fragile, a blend of Mourvèdre and Grenache, has firm herbal notes with fresh sappy fruit on the palate. The Grenache gives colour and the Mourvèdre some savoury notes.

The first red wine from Domaine de la Pertuisane is the wonderfully named Green Eggs and Vin (a reference to the Dr Seuss book *Green Eggs and Ham*). It is an extraordinary mixture of grape varieties, a field blend, with white grapes accounting for one-third of the wine, so with apologies for the list: Grenache Noir, which accounts for 60 per cent of the blend, Carignan, Alicante, and the whites Grenache Gris, Macabeo,

Grenache Blanc and Carignan Blanc, from two vineyards totalling 60 ares. Richard does not know how old the vineyards are, and they produce just two barrels. He picks the red grapes about ten days after the white, and co-ferments. The colour is surprisingly deep, given the quantity of white grapes, and the oak is quite obvious, but on the palate there is some intriguing spice – I thought of *pain d'epice* – and the flavours are ripe but fresh. Richard uses natural yeast and retains some stems to add a bit of grip to the wine, which is not filtered.

Le Nain Violet is a pure Grenache Noir, from one vineyard planted before 1950 and another, smaller, vineyard planted in the 1990s, all on schist. The wine is given a long maceration and spends 15 months in oak, including some new oak. The colour is deep and the wine rich and intense, with black chocolate, cherry liqueur and fruit, and silky tannins. For Richard, the key to Roussillon is old Grenache. Its yields are low, 12–14 hectolitres per hectare, as *coulure* can often be a problem.

Final Draft comes from just 20 ares of Cabernet Sauvignon planted some 15 years ago, which produces about 550 bottles. It spends 15 months in wood, one new barrel and one old barrel. The wine is rich and opulent with sweet blackcurrant fruit. The taste is certainly more reminiscent of the Napa Valley than Bordeaux. To Richard's knowledge his is the only estate in Roussillon to do something serious with Cabernet Sauvignon.

The red wine tasting of D66 began with Others, which is a very apt name, as it is made up of 60 per cent Grenache with the remaining 40 per cent made up of everything else. All fruit is handpicked and destemmed and given a week of cold soaking, followed by two to three weeks maceration, and then a five-day post-fermentation maceration. Ten months in oak follow. The colour is very deep and the wine ripe and rounded, with sweet, chocolate fruit. As Richard observed it is very much a Napa Valley interpretation of Roussillon, describing it as an iron fist in a velvet glove. Dave visits regularly throughout the year, and is there at harvest time. Richard noted their different approaches: he picks earlier and uses more stems, whereas Dave favours meticulous sorting and wants purity of fruit. Richard ages his wine for longer, between 15 and 18 months, rather than Dave's ten months.

D66 is a blend of 82 per cent Grenache Noir, balanced with 18 per cent Syrah, blended as late as possible, after ten months in oak. The colour is very deep, with some firm tannin balancing sweet fruit. Richard is planting more vines for D66, observing that if you keep the crop light for young vines, you can make good wine. Contrary to local popular belief, Carignan does not need to come from old vines to be good.

Domaine Pouderoux

Maury

www.domainepouderoux.fr

Robert Pouderoux trained as a lawyer but always wanted to be a wine grower. His parents had rented out his grandfather's vineyards, but he has gradually recuperated them, some 20 hectares in Maury, St Paul-de-Fenouillet and Latour-de-France, six plots spread out over about 12 kilometres. He started gradually, bottling his first wine in the early 1990s, converted to organic viticulture about 15 years ago and now follows some of the principles of biodynamic viticulture. When he started out, he only made Vin Doux Naturel; unusually for these days, Vin Doux Naturel still accounts for 40 per cent of his production.

We tasted in Robert's neat little tasting caveau, with the barrel cellar next door. His pure Vermentino Côtes Catalanes is fermented in old demi-muids and the oak was well-integrated. Robert likes Vermentino, finding that it has good acidity, more than Grenache Blanc. There were saline notes on the finish. He thinks that the white wines of Roussillon should be straight (or *droit*) and tight and certainly not too broad and fat.

A Côtes du Roussillon Villages from old Carignan, as well as some Syrah and Mourvèdre, grown on sandstone at Latour-de-France was nicely structured. Robert finds that Mourvèdre works well, as it is a late variety and less alcoholic than Syrah or Grenache. Grenache is, of course, the principal variety in his vineyards. The Maury Sec, Montpin, a *lieu-dit* and a *sélection parcellaire*, comes mainly from Grenache (planted by his grandfather in the 1960s) along with a little Carignan, which provided the tannin. It was sturdy, with structured oak. I quizzed Robert about the typicity of Maury. For him it is Grenache Noir on black schist. Most of the wines are very powerful, whereas he would like less power, and he favours cooler and higher vineyards, facing east or north, but never due south. He sold half of his original vines, as they were in full sun, and bought vineyards with shade, which were much less expensive. At the time, the cooperative was only making Vin Doux Naturel, and wanted vineyards that would give lower yields and higher alcohol. Today people are looking for cooler sites.

Robert considers that there are more differences between grape varieties than there are between vintages. He vinifies each variety separately to allow for blending opportunities and tries to maintain his own style, which he would describe as 'quite classic, not extravagant, and not concentrated, with *droiture* [directness or focus] and fruit'. His Maury Sec Terre Brune is just that, a blend of Grenache and Mourvèdre with elegant, fresh fruit and good structure. 'Mourvèdre calms the fire of the Grenache, restraining it by providing a counterbalance,' he says. He thinks that Mourvèdre is being rediscovered, whereas Syrah really suffers in the heat.

Robert makes three different Vins Doux Naturels, all from Grenache, which all begin in the same way. Vendange Grenat was given a cool pre-fermentation maceration, and muted *sur grains*, with 25 days maceration, before spending about eight months in vat. It was bottled in the spring following the harvest and is redolent of ripe fruit. This is the wine that made the reputation of the estate, whereas 30 years ago, in complete contrast, Maury was always made in an oxidative style. With Maury Mise Tardive Robert has copied a Late Bottled Vintage Port. The base wine is the same as the Grenat, but half of the wine spends two years in demi-muids, which are regularly topped up. It was a lovely glass of wine, more structured from the oak, but with fresh red fruit and elegant restraint. Initially this wine had problems with the *agrément* for the appellation: Maury was not meant to have an oaky taste. However, the oak in this instance adds complexity.

The Hors d'Age is kept for 15 years, in a solera of sorts, with barriques and demi-muids, so that the youngest wine is 15 years old. It was delicious, brick-red in colour with notes of chocolate and coffee and a nutty, elegant, long finish.

Clos del Rey

Maury

closdelrey@wanadoo.fr

It was a brilliantly sunny March morning, almost too good to be true, when Jacques Montagné took us into his vineyards in the hills south of the Agly Valley, towards the village of Rasiguères. We took a twisting track, on a magical mystery tour. Rosemary and thyme were in flower; there was a *cazot*, a small house, with almond and cypress trees (the olive trees died in the severe winter of 1956). The first pink Balearic cistus was just coming into flower; the white Montpellier cistus flowers a little later. There were views of the snow-capped Canigou, and you could see the tower of Trémoine, which overlooks Rasiguères. A pair of partridges scuttled across our path, diving into the garrigue of box, mastic, holm oak, juniper and honeysuckle. The wild plum was in flower, a delicate splash of white. There was wild asparagus waiting to be picked. Unfortunately, the area is paradise for wild boar so sturdy electric fences are essential. I learned how to tell the difference between Carignan and Grenache in a *complanté* vineyard: Carignan has redder stalks, Grenache vines are more yellow. Jacques explained that they delay pruning the Grenache so that it flowers when the weather is more stable, and hopefully will suffer less from *coulure*. They tidy up the vines in December, and then prune properly, leaving two eyes, at the end of March.

The first vintage of Clos del Rey was 2001. The name comes from the *lieu-dit*, Coume du Roy, but that name was already taken. The family have been wine growers for several generations: Jacques knows that his grandfather, who was born in Maury in 1883, was a vigneron. Jacques' son Julien makes the wine. Altogether they have 50 hectares of vines, and bottle wine from 12; the rest of their crop goes to the cooperative. When Julien finished his studies back in 2001, the cooperative was really only making Vin Doux Naturel, so that was the propitious moment to create their own estate.

In their functional cellar just outside the village of Maury cement vats have replaced the fibreglass ones. Their methods have also changed; they avoid what Jacques called excessive make up, in other words too much oak, and now prefer shorter macerations, with gentle *pigeage*. They have just four barriques, and would like a larger foudre.

We tasted in the cellar, with the *égrappoir* doubling up as a tasting table, and Jacques talked enthusiastically about their wines. First came a Côtes du Roussillon, La Sabina, based on Grenache with a little Syrah and Carignan. The name means a young juniper tree. There was a satisfying combination of ripe fruit and tension, with a certain stoniness balanced with tannin. L'Espicata, the name of the vineyard, was a pure Grenache with ripe, opulent fruit. L'Aragone, a pure Carignan, kept in vat, was more restrained with a firm streak of tannin, rich, red fruit and a fresh finish. Our tasting ended with Clos du Roy, a blend of equal parts Carignan and Grenache, of which 20 per cent is aged in oak, making for riper, sweeter flavours.

Domaine Semper

Maury

domaine.semper@wanadoo.fr

Mathieu Semper reminded me that *semper* is the Latin for always. His grandfather arrived in the region from Spain in 1930, at a time when there were a lot of Spaniards emigrating to Roussillon, married a local girl and planted a vineyard. Mathieu's parents, who retired in 2010, brought the family estate to 30 hectares. Mathieu is now the winemaker and his brother Florent runs the vineyards. Half their vines are in Lesquerde, and half in Maury. When Lesquerde was recognized as one of the specific Côtes du Roussillon Villages in 1977, Maury was still only producing Vin Doux Naturel. Their vineyard in Lesquerde previously belonged to Roger Mathieu, who was one of the very first to plant Syrah in the region, back in 1967. He brought cuttings, 2,000 vines from the Ardèche, and helped develop the village appellation. Lesquerde was not considered suitable for Vin Doux Naturel as it is a cooler area, with vineyards at 400 metres, but instead produces particularly good Carignan, grown on granite. Syrah was originally planted to improve the Carignan, and has adapted to the climate, benefiting from the higher altitude.

Talking to Mathieu, you sense that he is a perceptive winemaker; he is adamant that his wines must express the terroir, that is his philosophy. 'The winemaker must keep out of the limelight,' he says. Our tasting in their old cellar just off the main street in Maury began with Regain, a Côtes Catalanes made from 80 per cent Grenache Gris and Grenache Blanc, with some Carignan Blanc and Macabeo, from old vines planted on a north-facing slope. The wine spends six months in demi-muids and the flavours were quite rich and textured, with honey and acidity. Mathieu talked of the wine expressing 'mineralité solaire et cailloux chauds' – sunny minerality and hot stones.

Faniae, a Côtes du Roussillon Villages Lesquerde, comes from 50 per cent Carignan with some Syrah and Grenache, with a classic vinification, a 15-day maceration and an élevage in vat. 'The wine makes itself,' observed Mathieu. He is particularly enthusiastic about Carignan; it provides a sense of place and the granite soil of Lesquerde gives finesse to the wine, adding structure and tannin, while the Grenache adds flesh. There were notes of pepper and *garrigue*.

Voluptas also comes from Lesquerde, and is mainly Syrah, with some Grenache and Carignan. Mathieu finds that Syrah is easy in the cellar, but difficult in the vineyard; here it performs well on granite, as it does in the northern Rhône. There was fresh fruit, with peppery notes. Clos Florent is a Maury Sec, from 80 per cent Grenache with Carignan, vinified together, so that the slightly underripe Carignan adds freshness to the slightly overripe Grenache. It was redolent of liqueur cherries, with ripe, rounded fruit and spice, balanced with a fresh finish.

Maury Viatge, which means voyage or journey, but is meant to make you think 'vintage', a term they are not actually allowed to use, is muted at 78 grams of sugar per litre, so they work on the balance of ripe grapes and the right moment of mutage. There was ripe berry fruit on the nose and palate, but it was not too sweet and had a touch of spirit on the finish. Mathieu talked of his desire to dust, *dépoussiérer*, the image of Maury and Vin Doux Naturel in general. The sucrosity adds volume, but not weight, and it could certainly be a food wine. But when I asked him if the people of Maury, *les maurynois*, drink Maury, he replied: 'the old people, yes, but not the younger generation.' Tuilé Ange d'Or comes from a 20-year-old solera of Grenache Noir. The colour was tawny, with a nose of liquorice and red fruit, and hints of walnut, with a firm bite and a long, sweet finish. We were treated to a Rancio Sec from the barrel. It is pure Grenache Gris, with a firm, salty nose and taste. The wine is not muted and the barrel never topped up. Mathieu described it as the soul of the estate, with a solera that was begun in 2005 when a barrel of Regain was too powerful to be accepted as vin sec.

There were more barrels out the back, in what were once the old stables, built in 1860 at a time when horses were regularly used for work in the vineyard. Our tasting finished with a pure Macabeo. The wine had spent two years in vat, and then at least six in wood and would be added to a 20-year-old solera, to become Maury Ambré. It was rounded and honeyed, with notes of orange peel and a firm bite. Mathieu called it the spirit of Maury. He is a passionate defender of Macabeo, which lends itself to the oxidation needed for a mature Ambré, stating that 'oxidation like this is only possible with Macabeo.' The flavour lingered in my mouth as I made my way further down the main street to my next appointment, pausing to admire a mural of an open cellar door on the other side of the street portraying a mirror image of Mathieu's cellar.

Domaine des Soulanes

Tautavel

www.domaine-soulanes.com

I took a winding road through the vineyards, in the direction of Tautavel, to meet Daniel and Cathy Laffite. Soulanes is the term for a south-south-east facing slope, and the *lieu-dit* of Daniel's largest, 4-hectare, plot. Daniel is a maurynois, born and bred in Maury. After a brief career as a Jaguar car mechanic, he worked for the previous owner of the estate, Jean Pull, learning from him as well as doing various *stages*. Jean was already farming organically in 1992, and when he retired in 2002 he sold 17 of his 44 hectares to Mas Amiel and Daniel bought the rest, choosing the plots he wanted. Back in the 1990s, they made very little vin sec; the production was 90 per cent Vin Doux Naturel, of which a large part was sold *en vrac* to Martini. These days the balance has completely reversed, to 90 per cent vin sec. However, Daniel remarked gleefully that his *plus grande fierté*, one of his proudest moments, was selling some Maury to a wine shop in Portugal!

Our tasting began with Kaya Blanc, a blend of Grenache Gris and Grenache Blanc, with Carignan Gris and Carignan Blanc, fermented in oak, to give some firm structure with fresh, stony fruit. The Carignan adds acidity to the Grenache, and the schist of Maury also makes for a good level of acidity. Next came Anonyme, from equal parts of Grenache Gris and Carignan Gris, with the same vinification, but intriguingly different. Daniel observed that people are looking for Grenache Gris, but very little is planted. He much prefers the traditional varieties to the newcomers, the likes of Marsanne, Roussanne and Viognier, and he wants his white wine to age.

Cuvée Jean Pull, after the previous owner, is a blend of Grenache, Carignan, a little Syrah and a hint of Mourvèdre, fermented in a concrete vat. It was perfumed and spicy, perfect for easy drinking. Kaya Rouge is a pure Carignan, with fresh red berry fruit. Les Salines is a blend of 85 per cent Grenache, with some Carignan, aged partly in wood and partly in concrete. It had fresh fruit, what Daniel called *noyau de cerise* or cherry stone, which for him is the typicity of Maury. Daniel wants freshness in his wines and seeks to avoid the obvious warmth of Roussillon, tending to pick early and not over extract. However, for him it is the work in the vineyard that really matters.

The Vins Doux Naturels began with a Maury Grenat, a pure Grenache, with rich fruit and a streak of tannin. Daniel wants the tannins to slip in, *glisser*, behind the sugar, so the wine should not be too sweet. An extra 10 grams of sugar per litre can mask the tannins; it is all a question of the balance between sugar and alcohol. His Maury Hors d'Age comes from the 1998, 1999 and 2000 vintages and was bottled in 2017. It was rich and nutty with an elegant, lingering finish.

Domaine Thunevin-Calvet

Maury

www.thunevin-calvet.fr

Jean-Roger Calvet comes from Maury and is not at all connected to the bordelais négociant house of Calvet. However, his business partner, Jean-Luc Thunevin, is a bordelais. Together they created the estate of Thunevin-Calvet, based on vineyards from Jean-Roger's father, who was a member of the Maury cooperative. Jean-Roger, who studied at Montpellier, became interested in vin sec when Vin Doux Naturel was in crisis during the 1990s. He began experimenting with just 1.5 hectares while he was still a cooperative member, and as a *jeune agriculteur*, sought a financial partner. Jean-Luc, best known for Château Valendraud in St Emilion, came to visit, and as a *pied noir* from Algeria, was particularly taken with the Carignan. The result was the creation of their estate in 2001, with the removal of the last vines from the cooperative in 2004. They now have 55 hectares, and a smart new cellar in *pierre du Gard* was built on the edge of the village in 2008.

Grenache Noir accounts for 70 per cent of the vineyards, with some Carignan and Syrah, a little Mourvèdre and also some Grenache Blanc, Gris and Macabeo. Jean-Roger's wife, Marie, had family vines too, and Jean-Luc Thunevin and his wife, Murielle, have also bought vines, all in Maury, apart from one vineyard at Lesquerde. Virtually all the wines are vin sec, not Vin Doux Naturel. Jean-Roger's philosophy entails a lot of observation in the vineyard – he is passionate about the vine. He is registered for HVE and works organically in the vineyard, but not in the cellar, where he enjoys experimenting. He has a sorting table and ferments in cement vats, with hot or cold water for temperature control and a 15-day maceration, including some *pigeage*. The grapes may be destalked. As for wood, he began with barriques, but now much prefers demi-muids and also has some small foudres because 'you can lose the profile of a wine with barriques.' He thinks his winemaking has become more restrained over the years.

Cuvée Constance Rosé, named after his daughter, is a blend of Syrah grown on granite in Lesquerde and Grenache grown in Maury. It was a cheerful *rosé de piscine* (Jean-Roger observed that the consumer does not take rosé seriously). Cuvée Constance Blanc, a blend of equal parts Grenache Blanc and Macabeo, was fresh and pithy. The sunny side of the vines gives aroma, and the shady side acidity. L'Amourette Blanc, named after a plant also known as Maxima Briza, which is grown in the vineyard as a cover crop, is a pure Grenache Gris, including some old 60- to 100-year-old vines. Part of the wine is fermented in new wood, with Grenache Gris on schist giving minerality.

Cuvée Constance Rouge, a blend of Grenache with some Syrah and Carignan all kept in vat following a short maceration, has rounded, ripe fruit, and L'Amourette Rouge, a blend of Mourvèdre with 40 per cent Grenache, was fresh and elegant, with satisfying structure. Jean-Roger is working on achieving more elegant flavours with Mourvèdre and ideally would like to try a pure Mourvèdre. Les Dentelles is a Côtes du Roussillon Villages, with 50 per cent Grenache, 20 per cent Carignan and 30 per cent Syrah – you must have 30 per cent Mourvèdre or Syrah for a Côtes du Roussillon Villages, and there is not enough Grenache in the blend for a Maury. Such are the constraints of French wine legislation. Part of the wine is kept in cement in order to retain the fruit, and part for 18 months in old barrels, to soften the tannins. It was rich and rounded, with a firm streak of oak and leathery finish. Côtes du Roussillon Villages, Cuvée Hugo, named after Jean-Roger's son, is 70 per cent Grenache, with 30 per cent Syrah. The Grenache is aged in foudres and the Syrah in Burgundian *pièces*, for about 18 months. It was rich and ripe, with a firm streak of oak and some black fruit and liquorice.

Surprisingly, perhaps, Jean-Roger does not make Maury Sec: he does not agree with it. He would have liked to make a pure Grenache or a pure Carignan, as they are the traditional varieties of the region, and make the appellation of Maury quite different from the other appellations, so that it really stands apart from the Côtes du Roussillon Villages. It should be at the pinnacle of the pyramid, he believes, something different and more expensive. The bulk price is only €10 per hectolitre more than for a Vin Doux Naturel, and that is far too cheap.

Divae is a pure Syrah, though officially it includes some Carignan, from a vineyard of granite at Lesquerde, lying at 400 metres. A lot of Syrah was planted in the 1980s, not necessarily with the best clones, or in the best sites, but this granite site reflects the heat. The wine is kept in new Burgundian barrels for 18 months and is rich and chocolatey, heady and tannic. Les Trois Marie is Grenache, with just 20 per cent Syrah, aged in foudres. Coming from the oldest plots, with yields as low as 15 hectolitres per hectare, it is deep in colour, with rich, intense black fruit on both the nose and palate. Jean-Roger admitted that his aim was to make the best Grenache of the department.

Domaine la Toupie

Maury

www.domainelatoupie.com

Jérôme Collas explained that a *toupie* is a spinning top or gyroscope. He has an old one that belonged to his grandfather and then his father. For him it is a symbol of balance and stability; as a vigneron he starts with a blank sheet each year, finding balance in the seasons and looking for it in his wine, concentration balanced with elegance. Wine growing is a 'métier d'artisan', he says, and as an oenologist, he enjoys the attention to little details. You sense somebody who is bright, intelligent, and a thoughtful winemaker. Jérôme studied at Montpellier and began his life as a vigneron in California before working for the INAO and then coming to Roussillon to work for the cooperative in Tuchan. He left in 2011 and his first vintage in Maury was 2012. He now farms 12 hectares of vines, in five plots, with different soils: limestone, schist and *galets roulés.*

In 2018 he produced his first vintage of a pure Macabeo, a Côtes Catalanes, vinified in vat, with a fragrant nose and fresh, elegant fruit. Jérôme observed that the Catalan grape varieties are unknown. His second white Côtes Catalanes comes from Grenache Gris with some Macabeo; it sees some barrel ageing, making for satisfying weight with a salty finish. His white Côtes du Roussillon comes from Grenache Gris, Grenache Blanc and Macabeo and is vinified in wood and bottled the following spring. The oak in the young wine was still quite obvious: I preferred the Côtes Catalanes.

The 2018 vintage of Impro Libre, from Grenache with 25 per cent Syrah, is the first wine Jérôme made without sulphur; he pointed out that you need fantastic hygiene for unsulphured wines. It had fresh, elegant fruit. A Côtes Catalanes, Solo Grenache, was just that, very ripe with liqueur cherry fruit and a tannic streak. Pirouette, from 40 per cent Syrah vinified by carbonic maceration and 60 per cent Grenache vinified traditionally, had some fresh, peppery fruit, as well as Grenache cherry fruit. Quatuor Côtes du Roussillon Villages is a blend of Carignan, Syrah, Grenache and Mourvèdre, of which one-third is aged in 400-litre barrels, following a long maceration of 20–25 days. It has some satisfyingly spicey fruit. A Maury Sec, Sur 1 Fil Rouge is a blend of mostly Grenache with some Syrah and Mourvèdre, kept mainly in vat, with ripe fruit, spice and a silky finish. Côtes du Roussillon Volte Face, the result of a *vinification intégrale*, comes mainly from Syrah, with some

Grenache and Mourvèdre. It was solid and concentrated with ripe fruit and a firm tannic streak – *un peu* too much.

It paved the way to the Vins Doux Naturels, beginning with Maury Blanc Tertio. Jérôme explained how he developed his range of Maury, beginning with Maury Sec and then Grenat, and finally Maury Blanc. It comes from Macabeo and he does not make it every year. He mutes at 90–92 grams per litre of residual sugar, which is not too sweet, so that the wine tastes of fruit and alcohol. It is lightly honeyed. Of the 50-or-so wine growers in Maury, only about ten to twelve of them produce Maury Blanc. Au Gré d'Eole Maury Grenat is muted at 80–85 grams per litre residual sugar and has spicy cherry fruit, with some sweetness, balanced by fresh acidity and an elegant finish. Jérôme has developed a well-conceived range of wines and deserves to do well.

Les Vignerons de Maury

Maury

www.vigneronsdemaury.com

The cooperative of Maury was once virtually the sole wine producer of the village. Until recently the only independent producer was Mas Amiel, but there has been a marked growth in the number of independent wine growers in recent years. It is one of the oldest cooperatives of the department, founded in 1910, after the viticultural crisis of 1907, and has undergone various upheavals and changes over the years. When Maury flourished, selling base wine to the vermouth producer, Byrrh, in the nearby town of Thuir, there were 400 members in three cooperatives. The three merged in 1974. In 2000 the lone cooperative was responsible for 1,700 hectares. That figure has fallen to 850 hectares, with just 100 members, for Maury has inevitably been affected by the crisis with Vin Doux Naturel. In the 1980s the cooperative produced 45,000 hectolitres, including 5,000 hectolitres of vin sec; these days production has dropped to an annual average of 20,000 hectolitres, with just 4,000 hectolitres of Vin Doux Naturel.

Bernard Rouby was president of the syndicat when Maury Sec became an appellation, and has been involved with the cooperative for over 30 years, so he is eminently qualified to talk about Maury, in a quietly enthusiastic manner. He explained how the cooperative made its first Maury Sec in 2011. The previous director, Thierry Cazach, who arrived in Maury in 2001, and worked at the cooperative for nearly 20 years, was largely responsible for the development of vin sec. Bernard described him as a very good technician, by which he meant a good winemaker.

Despite its difficulties, the cooperative produces a sound range of wines that represents the two appellations of the village. Les Maurynates, a term for an inhabitant of Maury, is a range of single varietal IGPs; next comes Tradition Granit and Tradition Schists. L'Akmé is the highest point, the best wine. As yet the appellation of Maury Sec does not include a white wine so the white wines of the village are IGP, but Grenache Gris would be a suitable contender for an appellation, as demonstrated by Nature de Schist, a Côtes Catalanes Grenache Gris with delicate herbal fruit. Maury is also part of the extensive appellation of Muscat de Rivesaltes; a Maurynate Muscat Sec was fresh and grapey. I compared Tradition Schists, from Grenache with 20 per cent Carignan, with Tradition Granit, from 60 per cent Syrah with Grenache. The first was more leathery and the second more peppery. For Bernard, Maury Sec is best drunk young, and is not intended to be kept. Apparently the INAO commission for the appellation considered that oak ageing for wines produced from schist has a drying effect on the taste. Maury is therefore better aged in vat than in barrel, but the same would not be true of wines from neighbouring Tautavel produced from grapes grown on clay and limestone. My tasting of Maury Sec finished with L'Akmé, a *sélection parcellaire* from Grenache with 20 per cent Syrah, grown on schist. It was ripe and rounded with spicy fruit, weight and concentration, and no oak.

The Vin Doux Naturel provides a useful introduction to the appellation. Maury Blanc, mainly from Macabeo, is soft and biscuity. Maury Ambré, from white grapes Macabeo and Grenache Gris, is aged in foudres for several years: the 2010 vintage was bottled in 2018 and was lightly nutty. Grenat Pollen, a pure Grenache, *muté sur jus*, had been kept in vat. It was lighter in colour than the Grenache Récolte *sur grains*, also from Grenache and kept in vat. The *mutage sur grains* makes for more structure, body and tannin whereas the Pollen had fresher, lighter cherry fruit. The two wines made a fascinating comparison.

Tuilé Rancio, mainly from Grenache Noir, is aged for an average of six years in 60-hectolitre foudres in a solera system: the foudres are never completely emptied. The wine was brick-red in colour with red fruit, tannin and acidity on the palate. Tuilé Vieille Réserve, after 10 to 12 years in demi-muids, was elegant with rounded walnut fruit. A 1988 Maury Tuilé Chabert de Barbera, named after the last Cathar, who died at Quéribus in 1274, had an intense nose, a rich, nutty palate and elegant fruit. The wine is kept for about 30 years, mainly in foudres, before bottling. Maury Tuilé Cuvée du Centenaire was brick-red in colour with a concentrated nose, and a long nutty palate. The oldest wine included in the blend came from the 1990 vintage, so that it was fresher than the Chabert de Barbera. The demi-muids are not usually topped up – that is an essential part of the process of making a *rancio* – but they are kept inside, otherwise the angels' share is too high, with too much evaporation. Bernard observed that the angels are thirstier in Maury than in Banyuls, the difference between the humid sea air and drier inland mountain air.

ST PAUL-DE-FENOUILLET

St Paul-de-Fenouillet is the next village up the valley from Maury, and part of the appellation. Quite a few winemakers from outside the region have come to the village.

Domaine la Bancale

St Paul-de-Fenouillet

bastien.baillet@gmail.com

Bastien Baillet took us to see his vineyards just outside St Paul-de-Fenouillet. He has five hectares, scattered in six villages, mainly in St Paul, but also in Maury, Prugnanes, Lesquerde, Trilla and Tautavel, although he is selling the last as it is too far away, causing logistical complications. The differences between the vineyards are considerable, not just in soil, but also in temperature and altitude. In 2020 the harvest in Maury began on 15 August, but in Prugnanes, only 10 kilometres away, it started on 15 September.

I quizzed Bastien about the name of the estate. Bancale usually translates as wobbly or unsound, but in Catalan it can also refer to vines on terraces. Bastien explained that it described their early beginnings, the start of their adventure, when they didn't even have a pump and everything was artisanal in the extreme. He comes from Normandy and studied oenology in Montpellier before moving to Roussillon. He made his very first wine in 2014, and has gradually increased the vineyard holding. Vines are cheaper here than in neighbouring Languedoc and there are more old vines. 'They are part of our identity, even if they require a lot of manual work,' he says. For Bastien an old vine is one planted before 1950 and initially that was all he had, but now he has planted Grenache Gris. He would love some Carignan Gris, but realizes he will have to plant it himself. In 2016, he was able to stop working for other people, and in 2020 his wife, Céline, who had been working at Le Soula, joined him full time in the cellar.

Bastien has just moved into a new cellar, a simple industrial building on the edge of the village, with the luxury of space after the very cramped cellar on the ground floor of his house in the centre of the village. The street was so narrow that the dustcart could barely squeeze by. There are fibreglass tanks and a few barrels, and Bastien is gradually equipping the cellar with a new press and some cooling equipment.

Bastien makes two or three cuvées of both red and white wine, and in 2019 he also made a rosé. He had almost sold out of wine in bottle, so we tasted from vat and barrel. Fleuve Blanc is a blend mainly of Grenache Gris, with a little Grenache Blanc and some Macabeo, with some firm minerality. A pure Grenache Gris from a Stockinger barrel was focused and incisive. Chair Blanche, the other white wine, an unoaked blend of Macabeo and Grenache Gris, was tight and structured. A vat of Grenache and Syrah, fermented separately before undergoing malo together, forms the basis of Fleuve Rouge. The Grenache comes from 80-year-old vines, with a low yield. Bastien explained that he has given up oak élevage for his red wines, and also shortened his macerations.

There was just one bottle to taste, of Chair Blanche 2019, aged in vat, from a plot that Bastien bought in 2016. The wine had some exotic notes on the nose with a textured palate and firm acidity. Bastien occasionally uses sulphur at racking, after the malolactic fermentation, but not at bottling. You sense that he is a thoughtful winemaker, paying attention to detail, and he deserves to do well. He is full of ideas: he would like some olive and almond trees, and maybe to make some sparkling wine, perhaps even vinegar.

Domaine Benastra

St Paul-de-Fenouillet

06 42 62 06 70

Joseph Paillé comes from the Loire Valley, where his family have vineyards in Anjou, in the Coteaux du Layon. He and his ex-wife Wendy Wilson ran Domaine Pithon Paillé before deciding to come south, as Joseph already knew the region well from working with his uncle, Olivier Pithon. He found an estate for sale, with vines in the village of Trilla, at 500 metres, and at even higher altitude in neighbouring Felluns. He now has 19 hectares, with more vineyards in St Paul, Prugnanes and Maury, and his cellar is on the outskirts of St Paul, opposite the fire station. Although he has three business partners, Joseph is left very much to his own devices. Benastra, in Catalan, means the good star, that brings luck.

We tasted on a hot afternoon and the first wine, Le Flamant Rose Qui Pète, a pét nat from Syrah, was wonderfully refreshing, a combination of ripe strawberry fruit with fresh acidity. Joseph explained that he picks the Syrah before it is quite ripe, in order to save it from the wild boars, which are a serious problem since they are not really deterred by any wires, electric or otherwise. He bottles it before the end of fermentation, at 12.5% abv. Gneiss, a Côtes Catalanes blend of Macabeo, Grenache Blanc, Grenache Gris and Marsanne from the villages of Trilla and Felluns, is kept in 20-hectolitre foudres and 600-litre barrels. Joseph runs the wine off the moment it starts to smell of oak, observing that wood can overwhelm the acidity. It was nicely rounded.

Joseph reflected on the differences between Roussillon and the Loire Valley, where he made 22 different wines, with just two grape varieties, Chenin and Cabernet Franc. There are so many appellations in the Loire, and you must vinify within the appellation, whereas in Roussillon, there is a much greater choice of grape varieties, with permutations of blends, and also fewer appellations, he explains. Joseph enjoys blending and you sense that he is happy with the change of region. La Petite Soeur is a blend of Lledoner Pelut, Grenache, Syrah and Mourvèdre, made with whole bunch pressing and some maceration, resulting in some fresh cherry fruit. He also has vines at Salses that ripen two weeks earlier but ferments them all together, some overripe, some underripe, making a *millefeuille* in the vat, adding more nuances of flavour.

La Garrigue Côtes du Roussillon Villages is a blend of old Mourvèdre, planted 50-or-so years ago, with Grenache, Syrah and Carignan. The wine is made using whole bunches and is given an élevage of 12 months in foudres and demi-muids, making for spice and freshness. Joseph is particularly enthusiastic about Carignan, calling it his *cépage fétiche*. It can be difficult: the stems break in the wind, it is susceptible to disease and it ripens late, but it is well worth it, he says. He has a plot of 100-year-old Carignan vines in St Paul-de-Fenouillet. He gives the grapes a short maceration and the flavours are very elegant, with a streak of balancing tannins. Joseph is making a statement, with Carignan as his most expensive wine.

He is also trying his hand at Vin Doux Naturel, ageing one small barrel of Maury each year for an Hors d'Age. Our tasting finished with a Grenat Maury, with red fruit and liquorice. You cannot make a living from this, however, he pointed out. Joseph has plans, with land to re-plant up in the hills at Trilla, and is working on organic viticulture. He would also like to have a go at making beer.

Domaine Grier

St Paul-de-Fenouillet

www.domainegrier.com

Domaine Grier represents South African interest in Roussillon. In the 1980s, Jeff Grier developed Villiera in Stellenbosch, quickly establishing a reputation for sparkling wine, with the help of Jean-Louis Denois of Limoux fame. Twenty years later the family was looking to extend its vineyard holdings and considering possibilities outside South Africa, initially in an English-speaking country. However, when Jean-Louis told them about an estate for sale in the Agly Valley, they came to visit and simply had a *coup de foudre*. It is a 32-hectare property, including 22 hectares in one block between St Paul-de-Fenouillet and Caudiès-de-Fenouillèdes. In the mid-1970s the previous owner, Jean-Michel Pech, had cleared *garrigue* to plant vineyards. Realizing that he was not really a winemaker he sold his estate, although he continues to run the vineyards for the Griers. Raphaël Graugnard makes the wine and manages the estate for the absentee owners.

There were initial difficulties. Raphaël admitted that the Griers had found sales more complicated than expected. People who buy New World wine do not necessarily buy French wine and the Griers had anticipated selling one bottle of French wine for every ten South African bottles. This proved wildly optimistic: in reality the ratio was 1:100. The winemakers came from South Africa and did not know the region, and their wines did not respond to the market. Raphaël, who had been working for the cooperative in Tautavel, was brought in to sort things out and, quite simply, they started again from scratch. His first vintage was 2010 and since then he has revamped the wines and their labels.

The range is destined mainly for the South African market. Perhaps surprisingly, rosé accounts for nearly two-thirds of sales. Raphaël talked about the differences in the winemaking between South Africa and Roussillon, explaining that South Africa favours greater extraction, with a five or six-day maceration and grapes being pressed when the tannins arrive. He prefers to allow the tannins to come gently, so the maceration lasts four to five weeks, making for more supple tannins. South Africa favours a lower fermentation temperature of 20°C whereas French winemakers will let the temperature rise to 28°C or more so that the yeasts develop glycerol. The first year they made two different cuvées employing the contrasting methods, and France won. There was more initial fruit in the South African-method wine, but it did not seem so ripe; the South African method also entailed more sulphur.

Our tasting began with Grier Brut, a Blanc de Noirs and an original blend of 50 per cent Macabeo, 25 per cent Chardonnay and 25 per cent Carignan Noir. The Macabeo is quite neutral, the Chardonnay a little creamy and the Carignan adds some acidity. A pure Macabeo was rounded, with some acidity, while a Chardonnay was lightly buttery with 25 per cent of the wine fermented and aged in barrels. Half the bunches are picked when they are golden, and the other half when they are still green. Cuvée Alba, a white Côtes du Roussillon from 80 per cent Grenache Gris, with some Grenache Blanc, is fermented in wood, making a rich, leesy wine with a touch of oak. The malolactic fermentation helps integrate the oak. The rosé is a blend of 70 per cent Grenache and 30 per cent Carignan, with fresh fruit, firm acidity, and a salty note on the finish.

A pure Grenache Noir was lightly spicy; one can taste that it comes from a cooler area. Odyssea, a Côtes du Roussillon, is a blend of Syrah, Carignan and Grenache; the appellation Côtes du Roussillon Villages does not extend as far as the cooler vineyards of Caudiès. The wine is kept in vat and is rounded and fresh with a hint of liquorice. Côtes du Roussillon TP3, named after the small Renault van once used to transport the grapes that features on the label, is a blend of 70 per cent Lledoner Pelut and 30 per cent Syrah, with spice and liquorice. More specifically this wine is made without any sulphur, entailing meticulous hygiene. Raphaël adds yeast to the grapes at harvest, *des levures d'occupation*, to prevent the development of brettanomyces. Galamus, named after the nearby gorge, is a blend of equal parts Syrah, Grenache and Carignan aged for 12 months in 400-litre barrels (they have stopped using barriques): it was ripe and solid. The final wine was the 2014 Olympus, a Côtes du Roussillon Villages blend of 75 per cent Syrah and 25 per cent Grenache made from grapes bought in Maury and Lesquerde and aged for 12 months in barrel. The wine, which is not made every year, was developing some attractive cedary notes. The subsequent vintage,

2017, made from pure Carignan grown in Lesquerde was aged for 24 months in new wood. A barrel sample showed sweet fruit with a firm, tannin streak. Raphaël enthused about Carignan grown on granite at Lesquerde, observing that Carignan does not support mediocrity and is also susceptible to disease, but is well worth the trouble.

Mas Karolina

St Paul-de-Fenouillet

www.maskarolina.com

Caroline Bonville is bright and bubbly. She has a cellar right in the centre of the village, a building originally used for storing hay, where she welcomed us warmly. Caroline revealed that she comes from a family of wine growers – her parents were from Champagne, but moved to Bordeaux when she was just one. Although she studied in Bordeaux she could not afford vineyards there. A friend inspired her to look in Roussillon, where she found two hectares of schist in Maury, behind Mas Amiel. She now has 18 hectares, including vineyards in Lesquerde, with granite and more schist at Rasiguères. There are a lot of old vines, and she is gradually converting to organic viticulture and has not used any weed killer since 2010.

Our tasting began with a Côtes Catalanes, a blend of 70 per cent Grenache Gris and Grenache Blanc, with some Carignan Blanc and Macabeo. Some of the Grenache Gris and Blanc is aged in barrel, with the Carignan and Macabeo in vat. The nose was quite firm, with a fresh, savoury palate. Côtes Catalanes L'Enverre, from old Macabeo aged in 500-litre barrels for 12 months, had firm stony fruit. It was very incisive and tight-knit with a saline finish. Caroline pointed out that Maury is not always an easy terroir as you need to pick early.

A Côtes Catalanes from Grenache, with 20 per cent each of Syrah and Carignan, grown at Rasiguères and aged in vat, was wonderfully drinkable, with fresh cherry fruit. Caroline enthused about the 2019 vintage. Her Côtes du Roussillon Villages comes from Maury, but cannot be a Maury Sec as the 50 per cent Grenache in the blend is not enough to meet requirements. It is blended with Syrah from Lesquerde and Rasiguères, and given ten months of élevage in barrique. The oak was well integrated, with *garrigue* and spice on the palate. L'Enverre Rouge, with 75 per cent old Carignan from Rasiguères and 25 per cent old Grenache from Maury, is aged in 500- and 300-litre barrels, some of which are new. Again, the oak was well integrated, with spicy red fruit and elegant tannins. Vertige is a Côtes Catalanes and a pure Syrah, from Lesquerde, aged for 18 months in new oak. It is not made every year; the 2016 vintage was redolent of oak and tapenade on the palate, when it was three years old.

Muscat de Rivesaltes had been aged in barrel for between eight and ten months to temper the ebullient Muscat character, making for an original interpretation that was sweet and smooth with notes of dried apricot. L'Irrésistible is a Maury Blanc from old Macabeo, from a solera of barrels kept outside. Caroline dampens the barrels if the weather is too hot. The colour is amber and the nose firm and nutty, with honeyed concentration and hints of orange peel on the palate. Maury Grenat, muted at 80 grams per litre of residual sugar, spends 12 months in barrel and has some fresh spiciness, balanced with a streak of tannin. Caroline wants to be able to taste the tannin as she believes that it makes the wine more digestible.

Quizzed about the typicity of her wines, Caroline replied that she wants to retain elegance and freshness. She has some of the most entertaining wine labels I have ever encountered, which suit her personality.

Le Mas de la Lune

St Paul-de-Fenouillet

www.lemasdelalune.com

Matthieu and Vanessa Courtay are newcomers to winemaking and to St Paul-de-Fenouillet. Matthieu was brought up in Paris, but spent time with his grandparents in the countryside of the Dordogne. Thanks to his father living in Toulouse he met Patrice Lescarret from Domaine Causse-Marines in Gaillac and helped him with the harvest during his student holidays. He then spent 20 years in the film industry, working his way up the ladder to become a director, but the travelling and prolonged absences were not good for home life, and he wanted to do something else – why not make wine? In an attempt to discourage him Patrice suggested working with him in Gaillac as a WWOOFer – WWOOF stands for worldwide opportunity on organic farms, an organization that offers work opportunities, usually for students – but as Matthieu explained, it didn't work. Vanessa too wanted to leave Paris and so they explored various regions. They love Corsica but it is impossible to buy land on the island if you do not belong there. Then they came to Roussillon on holiday and fell in love with the area; it is wild, not unlike Corsica. They bought a cellar, house and vines, but only the old vines, from a bordelais who was selling up, and then as Matthieu put it, they made eyes at their bank manager, who was receptive. So Le Mas de la Lune was created in 2017 with six hectares of old vines in one large plot on black schist.

Matthieu and Vanessa are absolutely committed to biodynamic viticulture. They work with the moon and the planetary calendar, hence the name of the estate. The moon affects the tides and according to biodynamic principles it affects liquids and is therefore fundamental to cellar work. It is a question of understanding the moon's energy. They use various biodynamic treatments, infusions of plants. They do not have a tractor. The vines they bought had been chemically treated and need to be restored to health. It is hard work, but talking with them, you sense they are very happy with their new life, doing something that they love. It is, they say, 'un projet de vie, un projet de famille'.

Our tasting, happily on a fruit day, which is the best day in the biodynamic calendar for tasting, began with their white wine, a blend of Grenache Gris with some Macabeo, with a cool, slow fermentation in tank followed by some ageing in old neutral barrels. The wine was fresh, elegant and lightly aromatic. They also make La Petite Lune, from Grenache Gris. Oh la la … is a cheerful rosé, from Syrah with some Grenache Noir, pressed, with the juice kept on the lees of the white wine, and then given a few weeks in barrel, making for rounded, ripe fruit. Envole-Moi, with a witty design of flying elephants on the label, is a blend of 60 per cent Grenache, 35 per cent Syrah and just 5 per cent Mourvèdre kept in vat, with some ripe fruit and a fresh finish. It is Vin de France, with Matthieu observing that €15 per bottle is a price barrier for this wine designation. Above €15, it is better to have the appellation. He would like an appellation for the village of St Paul-de-Fenouillet. They are planning another red cuvée, with some oak ageing, which will be Côtes du Roussillon Villages, based on Grenache, with Carignan, Mourvèdre and Syrah.

Asked about future projects, Matthieu said, the aim is, quite simply, to continue to develop. He has new cuvées in mind, including a 12-day maceration of Macabeo in 2020, and he enthused about the excitement of following the vines as they grow, and the pleasure in getting up at 5 o'clock on a summer morning to go out into the vineyard. We went away with a bottle of their very first wine, 2017 OxyLune, from Grenache Gris and Macabeo, which spent 24 months in oak. It was very intriguing, quite strongly marked by the oak but with tannin and acidity and some dry fruit. Matthieu has made it again in 2020.

ST MARTIN-DE-FENOUILLET

From St Paul de Fenouillet, it is a short drive up into the hills to St Martin, passing the dramatic scenery of the Clue de la Fou, with the road weaving round every bend.

Domaine Eric Laguerre

St Martin-de-Fenouillet

www.domainelaguerre.com

Wine cellars can be dangerous places, as Eric Laguerre found out to his cost when he slipped in his cellar and concussed himself on the day in March 2020 when I had arranged to visited him. Instead we met up in late July at his house in St Martin-de-Fenouillet, where he was convalescing, stoically enduring his physical limitations. I remember meeting Eric on my first venture into the hills of the Fenouillèdes. Initially he worked in his grandfather's cellar and then with Gérard Gauby and at Le Soula for a couple of years, but then decided to develop the family vineyards, with a first vintage in 2004. Altogether he has 40 hectares, all around the village of St Martin, but is inevitably considering how many to keep and may just retain the best 15 hectares.

I encouraged Eric to talk about the changes he has seen over the last 30 years. The first thing he mentioned was the lack of water. It used to snow. He remembered one year when a metre of snow stopped him going to school for a week. The rain was more regular too, but it has always been hot. However, the streams have disappeared, and the rain has become more violent, with fierce storms. Eric talked about life in the countryside, what he called the *monde paysan*; he remembers fetching milk that was still warm, straight from the cow. His father never used weedkiller and hardly any chemicals; he had horses, mules and cows, and worked the soil. St Martin, which is one of the cooler villages of the Agly Valley, has always been a *terroir de blanc*, best for white wine, which both his grandfather and father made, selling it to Vignerons Catalans. His grandfather bottled wine as early as the 1960s. And 30 years ago, they used to deacidify the red wine at the village cooperative – yields were higher and the grapes less ripe. Carignan at 11% abv does it a gross disservice. It was of course blended with a richer wine and then sold on to the *négoce*. Grapes are much riper these days.

As we talked, we enjoyed Eric's wines. Eos Blanc is a blend of equal parts of Grenache Blanc and Macabeo, grown at 500 metres. The wine is nicely textured with good balancing acidity after eight months in demi-muids. Le Ciste is a blend of Grenache Blanc and Grenache Gris, with a little Marsanne, Roussanne, Vermentino and Macabeo, given 12 months in barrel, including a little new oak. Eric's father planted the Marsanne, Roussanne and Vermentino. There was a satisfying depth of flavour on the palate, fully vindicating the reputation of St Martin for white wine. Eric observed that he does as little as possible in the cellar; he is patient and believes that if you have good grapes you cannot get bad wine. There is no *bâtonnage* and he never racks. He adds a little sulphur after the malolactic fermentation and at bottling. Good hygiene is essential.

Eos Rouge is a pure Grenache Côtes du Roussillon, for St Martin was considered too cool to be classified as Côtes du Roussillon Villages. Nor is it within the appellation of Rivesaltes as it is too high in altitude. Eric remains loyal to the appellation out of respect for the work his father and grandfather put in to establishing it. Eos has lovely ripe fruit, with a streak of tannin and a fresh finish, making a beautiful expression of granite-grown Grenache. Le Ciste is more complicated, a blend of Syrah, Grenache, Carignan and Mourvèdre, aged in 25-hectolitre foudres for 12 to 18 months. It was rounded, with red and black fruit, and the freshness that is the hallmark of St Martin.

The final *bonne bouche* was Eric's interpretation of *rancio,* a Vin de France that was pure Grenache Blanc. He first made it in 2006. The wine is kept in 600-litre demi-muids for four or five years, in a cellar that is no warmer than 20°C and over the years there is a gentle evaporation. The wine was firm and nutty, with a dry finish, but not as severe as some *rancio.* I left, full of admiration for Eric's courage and determination; his fortitude is remarkable.

Domaine le Soula

St Martin-de-Fenouillet

www.le-soula.com

Domaine le Soula was created in 2001, when Gérard Gauby suggested to Mark Walford of the pioneering wine importers Richards Walford: 'we should invest'. Gérard felt that more should be done to save some of the wonderful old vines in the hills around the villages of St Martin-de-Fenouillet and St Arnac. Since 2001 there has been a series of talented winemakers at Le Soula, most of whom have gone on to create their own wine estates in the region: Tom Lubbe was followed by Jean-Philippe

Padié and then Eric Laguerre. Next came Vincent Balança, who left to make wine in Faugères, followed by Gerald Standley, a South African who arrived in Roussillon from Bordeaux. Gerald has now returned to Bordeaux, to be replaced by Wendy Wilson. She too is South African-born, but has had a varied career in wine, mainly in Virginia and then the Loire Valley before coming to Roussillon with her French husband, Joseph Paillé, who is related to the Pithon family. Wendy and Joseph have now gone their separate ways, and he has created his own small estate, Domaine Benastra, in St Paul-de-Fenouillet (see page 108).

We met at their offices and tractor shed in the hills outside St Martin-de-Fenouillet. It is an elegant construction built of local wood, and the vineyards are close by, while the vinification cellars are outside the village of Prugnanes. We went to see the vineyards, situated at altitudes between 420 and 650 metres, and Wendy talked about biodiversity. Sheep had been grazing in the vineyards earlier in the year and there are bee hives too, although the Asian hornet is doing great harm to bees in the area. It was a wonderful spring morning, with views of the Canigou in the distance and I spotted a hoopoe flitting through the trees. Wendy talked about *la taille douce*, soft pruning, whereby the vines are cut following the flow of the sap so that there are no hard angles. She lamented the damage caused by wild boar: they lost 100 hectolitres to boar last year. The vineyards are protected by electric wires, but they really need something more permanent. In the middle of March, they had not yet finished pruning, as they are concerned about the high frost risk. There are old vines, some Macabeo planted in 1900 and Carignan in 1919, with 16 plots making up 22 hectares. The soil is mainly granite, with some limestone.

Le Soula is the *lieu-dit*, where they have a plot that faces due south. The name means a sunny slope in Catalan. Wendy was intrigued by the fact that one of India's leading wineries is called Sula and an Indian friend had offered the explanation that both Hindi and Catalan originate from Sanskrit, and the Sanskrit word for sun is 'sula'. The wines are virtually all either Côtes Catalanes, with Wendy considering the IGP to carry more weight than an appellation, or Vin de France. The range comprises Le Soula red and white, Trigone red and white, a rosé and an orange wine, Macération du Soula Blanc. There is a flexibility about vintages and blends; some wines may well be a blend of vintages, as well as varieties.

Trigone Edition no 18, Vin de France, a pure Syrah from the 2018 vintage, was fresh and elegant with an emphasis on drinkability, and a refreshing 12.5% abv. Wendy wanted to show that Roussillon is capable of making wines that are ripe, while avoiding high alcohol. In contrast Trigone Edition 2019 is a Vin de France from three vintages and varietals, namely Syrah, Carignan and Grenache. It was perfumed and lightly peppery, with elegant structure, and very drinkable. Wendy favours what she calls a restrained maceration of 10 to 15 days, with whole clusters and no pumping over. All the wines spend time in oak, some Stockinger foudres of 20–30 hectolitres and some demi-muids and barriques from the bordelais cooper Boutes.

In the spring of 2020, the current vintage of Le Soula red was 2013, and they hold back even older vintages which are released when they are ready for drinking. Wendy opened the 2009, which is a blend of 70 per cent Grenache Noir, with Syrah and Carignan. It had intriguing leathery notes, with some silky tannins and was warm, without being heavy.

They only began making rosé in 2016. It was Wendy's idea – Mark's reaction was, 'Who is going to drink that?'. Many more people than he'd expected, it turns out. The 2019 vintage has some colour; you cannot avoid colour and it is a food rosé, suitable for all seasons, not just summer. It comes from four plots of Syrah, from pressed grapes, and spent five months in neutral oak to yield a wine with rounded raspberry fruit, some vinosity and a dry finish. I would be interested to see how it ages.

Trigone Blanc No 19 is a Vin de France and a blend of 90 per cent Macabeo, from both the 2019 and 2018 vintages, with 10 per cent Tourbat from 2018. The grapes are pressed, but not destemmed, and the juice goes into barrel until it is ready for bottling. The wine was rounded, with good acidity and some texture in the mouth. The current vintage of Le Soula Blanc is 2016, which Wendy has released before the 2015, as she felt that it was ready first. It was indeed drinking beautifully, with rounded fruit and good acidity. The blend of grape varieties is certainly unusual: equal parts of Grenache Blanc, Macabeo and Sauvignon Blanc, with a little Marsanne, Roussanne, Vermentino and Grenache Gris. All except the Marsanne and Roussanne are field blends, and the wine spends 18 to 24 months in barrel. Wendy asserted that there is no recipe: she begins the harvest with the Sauvignon, and ultimately, she would like to make a pure Sauvignon Blanc. It performs well on granite soil at a high altitude, producing good acidity.

No 19 Macération du Soula, an orange wine, comes from five vintages, between 2014 and 2019, and the grape variety mix is 60 per cent Vermentino with 40 per cent Macabeo. Whole clusters are fermented with a 10- to 15-day maceration. The juice is tasted frequently during the fermentation: when you start getting aroma, you also start getting tannin and that is the moment to press, explained Wendy. The wine is then aged in barrel. I thought it was delicious, with an orange nose and a rounded palate, depth and texture, and a dry finish with a streak of tannin. Wendy enthused about the versatility of orange wine. She is thrilled to have settled in Roussillon, describing it as a hidden gem and finding it very exciting. Her enthusiasm is infectious.

ST ARNAC

Domaine Paul Meunier Centernach

St Arnac

www.paulmeunier-centernach.com

Paul Meunier comes from Burgundy; his parents have Domaine J.J. Confuron at Premeaux-Prissey near Nuits-St-Georges, which his young brother will eventually take over. Paul is fascinated by Port, which led him to Maury, working for six months as a *stagiaire* in 2009, and returning regularly for holidays. Then he found his first vines and in 2014 took over the cellars of the Préceptorie de Centernach, which had been formed by a group of ten former members of the cooperative of Maury, and was financed by the Parcé family. When the Parcé family was no longer interested, the Préceptorie came to its natural end and the cellars in St Arnac, a tiny village in the hills above Maury, became available. We joked, with Paul explaining that Arnac is a corruption of the Catalan name of Centernach; it is not a saint's name but since *arnaque* is the French translation for the film *The Sting*, maybe St Arnac is the patron saint of gangsters or financiers.

Paul and his *compagne*, Lucille, have ten hectares of vines over ten plots in the nearby villages of Maury and Lesquerde, planted with Grenache, Lledoner Pelut, Carignan and Syrah, and for white wines, mainly Macabeo, as well as some Grenache Gris and Grenache Blanc, and a little Muscat à Petits Grains. Five hectares of his vines have very narrow rows, where using a tractor would be impossible. His sister, Périne, helps for six months of the year, working as a *prestateur* with a horse. Until the severe winter of 1956, St Arnac was known for its olive trees, but many were killed by the extreme cold. There are very few vines in the village itself. Paul and Lucille are planting olive trees, as well as pistachios and almonds.

Paul's welcome is very Burgundian. We tasted samples from vat and barrel, first 2018s and then 2017s, with Paul commenting on the various wines. There was a Grenache Noir *complanté* with some Macabeo from Maury. A pure Carignan was dark and spicy, grown in vineyards at 400 metres. The red wines spend two winters in barrel and all the grapes, except the Syrah, are destemmed. Paul told me that his overheads are much lower than they would be in Burgundy. A Syrah from Lesquerde, from 40-year-old vines, was fresh and peppery with some tannin; in contrast, there is very little Syrah in Maury, thanks to *esca*, and the effects of drought. These will be the future Côtes du Roussillon and Côtes du Roussillon Villages, as well as a *sélection parcellaire*, with Paul preferring to make Côtes du Roussillon Villages, rather than Maury Sec. He talked about the soil; Lesquerde is granite, but everything was squashed together when the Pyrenees were formed, making 'un gros bazar' meaning that the soil can change quite dramatically from one village to another.

Paul is lucky in having a very reliable source of old barrels from his father. The white wine, a Côtes du Roussillon, is fermented partly in amphorae from Spain, and partly in vat. Half the wine is given 24 hours' skin contact, and the wine undergoes a malolactic fermentation. The young wine tasted firmly salty, with good acidity and some satisfying mouthfeel.

Next came the Vins Doux Naturels, an elegantly grapey Muscat de Rivesaltes followed by a young Maury from Grenache Noir, that was redolent of liqueur cherries. A Maury Rancio was tawny in colour and intriguingly nutty, with a firm bite on the finish. Paul observed that cellars in Roussillon are traditionally dry and the resulting evaporation makes for higher alcohol and greater concentration in the wine. A wine of 17% abv can easily rise to 20% abv over time.

Quizzed about differences between Burgundy and Roussillon, Paul highlighted the Burgundian emphasis on taste, rather than technique. His methods haven't changed since his first solo vintage in 2014. He gives his red wines a 15-day maceration, but no pre- or post-fermentation, and prefers *pigeage* for Grenache and *remontage* for the Syrah. As for the Carignan, that extracts very nicely without any extra help. He is still looking for his style for his Côtes du Roussillon Villages and would like more freshness. Paul admitted that initially he had not wanted to be a wine grower. He has worked in New Zealand, at Rippon Vineyard in Wanaka, and also in Australia, Chile, Lebanon and Portugal, hence his enthusiasm for Port. Paul lamented that Vin Doux Naturel is so under-appreciated, with the local rugby players preferring to drink whisky and pastis, not wine.

Our conversation continued over a friendly lunch, *en famille*, with more bottles, and I left with a long, lingering taste of 2015 Maury Grenat, La Serrelongue, which went extraordinarily well with a *crème catalane* with a hint of lemon.

7

ESTAGEL TO CALCE

The named villages of Côtes du Roussillon Villages have enjoyed mixed fortunes. From St Arnac, you go back towards the valley floor and reach Lesquerde. I visited several wine growers with vineyards in that commune, but no one with a cellar there. The terroir is granite, which is good for Syrah, but too cool for Vin Doux Naturel. The village cooperative has joined up with that of Estagel so that any wine from Lesquerde is made in Estagel. From Lesquerde, the road leads to Maury, and then continuing east you reach the little town of Estagel, where there are several wine growers and the aforementioned cooperative, the wines of which go under the label of Les Vignerons des Côtes d'Agly.

ESTAGEL

Domaine Raymond Manchon – Bota Nostra

Estagel

04 68 29 17 11

We tracked down Raymond Manchon in the narrow back streets of Estagel, at the top of a hill right on the edge of the village, with a plot of vines just outside his house. Raymond is nicely energetic and opinionated. His family has been in Estagel since the seventeenth century, which makes him a true estagellois. He speaks Catalan and married a girl from the village of Tuchan in the Corbières, on other side of the hills, who speaks Occitan. The family were *ouvriers agricoles*, farm workers, working the land with a few vines of their own; one of them was also a *sabotier*, who repaired rather than made clogs. His father had 7.5 hectares of vines, which Raymond eventually took over, making his first wine in

2008, though he had joined the Estagel cooperative in 1993. He now has 20 hectares, which have been farmed organically since 2009. The production of a significant amount of his vineyards goes to the cooperative, Les Vignerons des Côtes d'Agly, which, he assured us, is the biggest producer of organic wine in the whole of the department.

Raymond describes himself as a fervent militant of Carignan, defending the identity of the variety and criticizing people who have tried to improve the appellation wines with Syrah, when Carignan is so much better. 'We must do what we do best,' he says. 'There are very few places where Carignan ripens so well as the Agly Valley. It is difficult if the terrain is too rich, as it overproduces and ripens badly. Old vines are great, but young vines do well if they are planted in the right place.'

Raymond did some agricultural studies, so that he could understand the soil, and he has worked at the Estagel cooperative, but essentially he has learnt his winemaking by asking others what they do. He uses as little sulphur as possible, and no cultured yeast, filtering or fining. It is all as natural as possible. Raymond enjoys the notion of liberty with natural wine, escaping from the *cahier des charges*, but notes the permanent risk of encountering a problem. The standard practice has become a lower alcoholic degree, a short maceration and an early bottling. Raymond, on the other hand, prefers a three if not four-week maceration, and even on one occasion a six-week maceration, determined by the tannins. He does not want to impose a technical process.

Cabrila is a pure Grenache Blanc, vinified in old wood for some micro-oxygenation, with firm, incisive fruit and good tension. Next was a pure Carignan, aged in barrel, with a touch of reduction on the nose, but with some appealing fruit and structure on the palate. It came from the 10 ares vineyard in front of the house. Another Carignan from a vineyard planted by his grandfather in 1926 had some richness and spice, balancing acidity and tannin. Jeff comes from another plot of Carignan with 15 per cent Grenache, both Noir and Blanc, as well as some Lledoner Pelut and Macabeo. The finale was Intrusio, a pure Syrah, rich, heady and concentrated. The name tells you what Raymond really thinks about Syrah. His wines are sold under the brand name of Bota Nostra, meaning our barrel.

Domaine des Schistes

Estagel

www.domainedesschistes.com

Jacques Sire created Domaine des Schistes in 1989 and it is now his son, Mickaël, who makes the wine. Although his name is Breton, he is pure Catalan. His great-grandfather was an independent wine grower who sold his wine *en vrac*, but when the cooperative in Estagel was founded in 1910, Mickaël's grandfather took the easy path. However, his father had always wanted to make his own wine, and his mother also had family vines, so they took the independent route over 30 years ago. The estate comprises 50 hectares in Estagel and Tautavel, in five large plots. Mickaël studied at Montpellier and has worked in Bergerac, the Côte d'Or, and further afield at Sherwood Estate in New Zealand and Darling Cellars in South Africa, which he described as '*une grosse boîte*'. His work experience taught him technical rigour and confirmed what he really wanted to do. His first vintage in Estagel was in 2004 and he has been in charge since 2009.

When asked how his methods differ from those of his father he told me he thinks his wines are fresher, with less obvious élevage in oak as he uses larger barrels. The whites have improved too, he says, and he has also made an orange Muscat. When his father began, half of his production was Vin Doux Naturel but these days it is just 250 hectolitres out of a total of 1,300 hectolitres.

Mickaël treated me to an extensive tasting in his neat little *caveau* in the back streets of Estagel. Illico is a pure Macabeo Côtes Catalanes, with fresh, herbal notes and sappy acidity. Mickaël observed that you need to pick Macabeo early to retain the acidity and avoid an excess of alcohol. Essencial Côtes du Roussillon is a field blend of Macabeo, Grenache Gris and Grenache Blanc, and also a blend of different terroirs of schist. There was no malolactic fermentation, but some *bâtonnage* gives some weight and fruit, with a satisfying mouthfeel. Casot d'Engora comes from a one-hectare plot of Grenache Blanc, with some Grenache Gris and Macabeo, given an élevage in small foudres for ten months.

Illico Rosé is a blend of one-third Syrah and two-thirds Grenache, with some raspberry fruit and quite a vinous finish. Mickaël admitted that he would like to make a more frivolous rosé and he has planted some Cinsault. Traditionally there is very little Cinsault in Roussillon as it is not suitable for Vin Doux Naturel. The yields are too low and it does not perform well on schist. Mickaël has planted it on sand and it will come into production in 2021.

Illico Rouge, a blend of Marselan, Carignan and a little Merlot is rounded and easy. Essencial Côtes du Roussillon Villages is a blend of one-third each of Carignan, Grenache and Lledoner Pelut, kept in vat, with ripe fruit and a firm streak of tannin, as well as acidity from the Carignan. Caume d'En Joffre comes from a plot in Tautavel of Carignan planted in 1963. The terroir of Tautavel gives more acidity than Maury, with blackcurrant rather than black cherry fruit. Tautavel has more Carignan in the vineyards there, as traditionally less Vin Doux Naturel was made there. Another single vineyard, Devant le Mas, planted in 1965 during a boom period for planting, is a Maury Sec, from Grenache and Lledoner Pelut grown on schist. The wine is kept in demi-muids for 12 months. It was fresh and elegant; with Mickaël pointing out that Grenache needs more oak than Carignan, which can lose its fruit in wood. La Coupeville comes from 40-year-old Syrah grown on schist with just 5 per cent Mourvèdre, given an élevage in small foudres; it has fresh, perfumed fruit and supple tannins. Mickaël feels strongly that his best wines must be an appellation, even if Vin de France allows for more liberty. He comes from Roussillon and is proud of the region and its wines.

Then we moved on to Vins Doux Naturels. Maury la Cerisaie comes from 60- or 70-year-old Grenache vines, and was redolent of cherry liqueur with fresh fruit. Muscat de Rivesaltes is a blend of Muscat à Petits Grains, with 20 per cent Muscat d'Alexandrie, which is usually the preferred Muscat for Muscat de Rivesaltes. However, for Mickaël, the higher percentage of Muscat à Petits Grains makes for more elegance. Joia, from two-thirds Muscat à Petits Grains and one-third Muscat d'Alexandrie is given a ten-day maceration. It was nicely characterful, with a refreshing bitterness and orange notes as well as a pleasing grapeyness. Mickaël thinks that Muscat can be made too technically, 'it is more like perfume than wine,' he says, and he wanted to make something more vinous.

A Rivesaltes from a solera begun by his father in 1990, a blend of all three colours of Grenache, had a dry, nutty palate and a lovely long finish. Algo, made for the first time in 2018, from Grenache Blanc and Grenache Gris, is a cross between a Rancio Sec and a white wine. It spends two or three years in a solera of sorts; there was no *flor* and it was not muted, so the flavour was very saline, with firm acidity. Its intensity made me think of a *vin jaune* from the Jura. The grand finale was a Rancio Sec, a Côtes Catalanes blend of Grenache Blanc and Grenache Gris, from a solera begun in 2004. The wine was orange-amber in colour and the taste firm and austere.

Mickaël is a thoughtful winemaker and an astute commentator on Roussillon. As for projects for the future, he would like to find some vineyards of Carignan Blanc and Carignan Gris, and plant some Tourbat. He was only 36 when we met in 2019 so his career will be fascinating to follow.

Domaine Vinci

Estagel
www.domaine-vinci.fr

We met Olivier Varichon at his house in Espira-de-l'Agly, where he stores most of his wine, rather than at his cellar in Estagel. The first thing you see when you go into the cellar is a splendid 1958 Renault, which he inherited from a wine grower in a village in Savoie, where his family owned the sparkling wine producer Varichon & Clerc. When his grandfather retired, the business was sold to Boisset. Olivier was not allowed to use his family name for his estate; Vinci is his wife's name. Emmanuelle comes from Roussillon and that is what brought them to the region. They had both worked in the wine trade in London but wanted to make their own wine and Olivier studied at Mâcon. They arrived in Roussillon in 2001 attracted by the inexpensive land and the old vines for sale.

We sat on the terrace in the shade of some wisteria on a hot summer's day. Olivier has seven hectares, in Espira, Cases-de-Pène, Opoul and Peyrestortes. In the spring of 2020 his vines were badly affected by mildew; the previous year they were sunburnt. Such are the vagaries of climate change. The heart of their vineyards is in the Combe de Rafalot, near Mas Crémat. Olivier has only local grape varieties, and no Syrah. The difficulties of finding a cellar meant that his very first vintage was made in a basement bedroom, but in 2004 they found an old warehouse in Estagel. He works organically, using natural yeast and minimal sulphur, looking for freshness combined with ripeness. He is working on the organic matter in his vineyards; animals graze there during the winter months and he is planting fruit trees, *pêches de vignes*, almonds and apricots.

Coyade, a pure Macabeo, is fresh with saline notes. Codals, meaning *galets roulés* in Catalan, from Grenache Gris and Grenache Blanc, has quite a full body after 12 months on lees. Roc is a pure Grenache, not just Noir, but also Gris and Blanc with some Lledoner Pelut, with ripe cherry fruit. Rafalot is a pure Carignan, grown in the Combe de

Rafalot. Early vintages were aged in oak, but now it is made in vat. Olivier laughingly described Carignan as *l'enfant terrible* of Roussillon. Maquis is a pure Lledoner Pelut aged in vat as it oxidizes too much in barrel. Olivier would also like to try amphorae, made from sandstone rather than terracotta. He is one of the many examples of outsiders coming to the region, and now has an English business partner, Simon Stoye, who wanted to buy his own vineyard but saw the wisdom of investing in an existing business.

LATOUR-DE-FRANCE

From Estagel it is a short drive back up into the hills to the village of Latour-de-France. This is one of the original named villages of the appellation, when the majority of the wine was produced by the cooperative and sold to the French wine merchant Nicolas. In normal years, the village hosts an annual open day. It has acquired a reputation for natural wines, *vins natures*, forming a magnet for many newcomers to the village. They are attracted not only by the reputation of the Agly Valley and the presence of other natural winemakers but also by vines for sale in interesting, relatively inexpensive terroirs. Since natural wines can usually be sold sooner than more conventional wines, finances are less constrained. However, as will be seen, not all the wine growers in the village are natural or even organic.

Domaine de l'Agly

Latour-de-France

www.skwine.fr

Boris Kovač suggested that we meet on the outskirts of the village, by the cemetery, so that he could guide me back to Domaine de l'Agly, which is in a stunning position in the hills above the village of Latour-de-France with views of Força Réal and Tautavel. With the lack of signs, it would certainly have been a challenge to find without a guide. Agly is apparently the old Catalan name for an *aigle*, an eagle.

Boris comes from Serbia, where his father was a well-known oenologist. Boris deserted from the Serbian army and came to France, studying oenology at Montpellier and initially working in a laboratory in Narbonne. However, he wanted the broader experience of making wine, so worked with Australian winemakers for the large organization

Foncalieu before spending some time in Australia. When he returned to France, he bought his vineyards outside Latour-de-France with two partners. Boris is responsible for the day-to-day running of their 12-hectare estate, which includes one hectare of white vines, Vermentino, Viognier and some Chardonnay, as well as Syrah, Grenache, Mourvèdre and a little Carignan. There are a lot of old vines, mostly field blends. The oldest are Carignan, planted in 1900, and the youngest were planted in 2006. In the vineyards Boris follows the principles of HVE, maintaining the *garrigue* and treating the vines only as necessary. He abandoned organic viticulture, owing to the problems with copper and the compacted soil from the increased use of a tractor.

Boris makes just four wines; two Côtes du Roussillon Villages, with the first vintage of a third wine not yet in bottle when I visited in June 2019, and one white Côtes Catalanes. Les Neiges de l'Agly is a blend of Vermentino and Viognier, so called because it snowed in 2006, the year the vines were planted. Snow is very unusual in Roussillon. The wine is lightly oaky, peachy and textured after two years in barrel. The vineyards are at 280 metres, with mica schist on black granite, which when combined with an acid soil gives good minerality to the wines. Yields are tiny, never more than 20 hectolitres per hectare.

Entrée du Royaume, the original name for the village Latour-de-France, on the border between France and Catalonia, is a blend of equal parts Syrah, Grenache and Carignan, with a small drop of Mourvèdre, given a minimum of two years in barrel, some new and some older. Boris is not looking for oakiness, but the better the barrel is, the longer he keeps the wine in it. This was rounded and rich, with supple tannins.

Les Cimes comes from the same plots, and the same varieties, but with the grapes left to hang longer on the vines. Boris enthused about blending barrels. Each grape variety is vinified separately in a small barrel, or maybe a tronconic vat. He takes the tops off the barrels and does a manual *pigeage*. With a long maceration, up to 45 days, Les Cimes has much more concentration and more alcohol, with intense cherry liqueur flavours. I admired it, but found it less elegant than Entrée du Royaume.

The final wine, Le Clos, is made from the plot of old Carignan, planted in 1900. The vineyard merits its name as it is on three terraces surrounded by a wall. The wine was still in vat, with fresh red fruit, and promised well, making a fine conclusion to a friendly visit.

Domaine de l'Ausseil

Latour-de-France

www.lausseil.com

Jacques de Chancel arrived in Latour-de-France in 2001 and bought his cellar in 2002. At that time there were hardly any other wine growers in the village besides the village cooperative, the very traditional Domaine de Rancy and Chapoutier, which had arrived in 1999. Jacques was born in Paris, to parents who had returned from Algeria in 1963. He studied agriculture and then went to Montpellier, specializing in oenology. After five years in the Médoc at Château Beaumont followed by a stint at Los Vascos in Chile, he began to consider doing his own thing, and looked at vineyards between Spain and the Gard. At Latour-de-France he liked what he saw, 'un joli terroir'.

Jacques makes an eclectic range of wine and opened several bottles as we chatted. His wines have eye-catching labels designed by his girlfriend. First was En Plein Vol, a pét nat from Mourvèdre. Next came a rosé, P'tit Piaf, in other words a sparrow, and a reference to Edith Piaf. Jacques talked about how his winemaking has evolved. He was not working organically when he started and used cultured yeast before he got to know the cellar, but he wanted to keep things simple, and has gradually evolved to use no cultured yeast or sulphur. Le Papillon is a blend of Macabeo and Grenache Gris with some firm fruit. P'tit Piaf Merlot was ripe and rounded, although Jacques explained that the variety was already in his vineyards and he is not a particular fan. Alouette comes from young vines, a blend of one-third Syrah, Grenache and Mourvèdre, all destalked. The Syrah and Grenache are picked together, while the later-ripening Mourvèdre is fermented separately and everything blended as soon as possible, making for good fruit and satisfying depth. Du Vent dans les Plumes, almost pure Carignan, save for a little Grenache, sees élevage in demi-muids and vat for 12 months. The Carignan comes from vines with an average age of 70 years and the flavours were quite firm and rustic. La Capitelle – he used to have a vineyard with a *capitelle* but no longer – is a blend of Carignan and Grenache, with some ripe cherry liqueur fruit.

The final bottle of vin sec was a 2010 Les Trois Pierres Côtes du Roussillon Villages Latour-de-France, the last wine that Jacques made with the appellation. A blend of Syrah from 50-year-old vines, with some Carignan and Grenache, it had some firm, blackcurrant gum fruit on the nose, and black olives and pepper on the palate. It had aged beautifully. We finished with a Rivesaltes Grenat from Grenache Noir *muté sur grains*, with a satisfying balance of fruit and sugar. Jacques suggested drinking it with a *magret de canard aux figues*. Suddenly it was lunchtime, and we adjourned to Le Coq à l'Ane, the wine bar in the centre of the village, which specializes in natural wine.

Domaine de Bila-Haut

Latour-de-France

www.chapoutier.com

Domaine de Bila-Haut belongs to the northern Rhône family of Chapoutier. Bought in 1999, it forms part of their extensive portfolio, including vineyards in Alsace, Germany, Beaujolais, Spain, Portugal and Australia. The vineyards of Domaine de Bila-Haut, 120 hectares in all, are mainly around the village of Latour-de-France, but there are also vineyards in Lesquerde along with a more recent purchase in cooler Tarerach, as well as an earlier purchase in Banyuls, mainly for vin sec, rather than Vin Doux Naturel. They are managed by Guilhem Domergue, who was a young boy when I first met him, when his parents were making ground-breaking Minervois in the 1980s.

Guilhem took us for an extensive drive through the vineyards, and talked with enthusiasm and perception about the differences between them. From a nearby viewpoint, the landscape is breathtaking, with the village of Latour-de-France at one's feet. Guilhem explained the different terroirs of the village. There is schist, which is good for Carignan and Grenache, but because it drains well Syrah can suffer from water stress in this very dry climate. The vineyards at Cassagnes are on gneiss, a kind of decomposed granite that makes for wines with fruit. The *lieu-dit* of Roubals is based on clay and limestone, on the banks of the Agly the soil is alluvial, which is better for Syrah, and at Lesquerde granite is the dominant soil.

The domaine works biodynamically, although in Banyuls they are not able to follow these methods consistently as the cost of labour there is so prohibitive. The vineyards at Latour-de-France are grouped on a large plateau, the previous owner having made a fortune in quarries. Grenache and Lledoner Pelut form the base of the red wines, with some Syrah and Mourvèdre, but less Carignan. The average age of the vines is about 25 years. White varieties are mainly Grenache Gris and Grenache Blanc. Apparently, Michel Chapoutier would like to plant Touriga Nacional.

As well as the estate wines, they have a négociant label, Les Vignes de Bila-Haut, and for each village, Collioure, Lesquerde and

Latour-de-France, there is a village wine and a special selection from the best plots. Wines from Les Vignes de Bila-Haut are blends of all four terroirs. The white, based on Grenache Blanc, with Roussanne, plus a little Vermentino, Macabeo and Marsanne, was rich but had good balancing acidity. The red wine comes from Grenache and Syrah with some Carignan and Mourvèdre and has ripe, spicy fruit with a fresh finish. Chrysopée is the selected plot from Collioure; the Syrah spends 18 months in demi-muids while the Grenache remains in vat, making for rich spice with a firm tannic streak. Occultum Lapidum from Latour-de-France is a blend of Grenache with some Syrah and a little Mourvèdre and Carignan, aged for 18 months. The vineyards at Roubals provide the selected plot at Latour-de-France, with Syrah and Grenache making for dense, ripe fruit. R.I. from Lesquerde is a pure Syrah aged in barrel, with some peppery fruit and a fresh finish. Essentially Guilhem wants to bring out the typicity of the soils without too much tannin or extraction, and he has succeeded. And what comes next? They are considering developing the white wines, and maybe adding a Banyuls Rimage that could age, plus there are the vineyards at Tarerach to consider.

Domaine de Rancy

Latour-de-France

www.domaine-rancy.com

Brigitte Verdaguer gave me a friendly welcome in her little tasting room in the heart of the village. She has a series of small cellars scattered around Latour-de-France where her Rivesaltes matures in old barrels. The aroma is intoxicating. This is the oldest wine estate in the village, with vines going back to 1919. In 1989 Brigitte and her husband, Jean-Hubert, took over the vineyards from his parents, initially following them in selling the wine *en vrac*. Then in 2001 they began producing red vin sec. These days their daughter Delphine is the winemaker and she is developing their range of vin sec, making the estate's very first white wine in 2017. However, they still retain the tradition for Vin Doux Naturel; Brigitte laughingly described them as 'the last of the Mohicans'.

They have 17 hectares. There are 12 hectares of white grapes, most of which is Macabeo although there is also some Grenache Gris for white wines and for the Rivesaltes Ambré. The five hectares of red vines comprise Carignan and Grenache, both nearly 100 years old, and Mourvèdre that is 35 years old. Yields average a meagre 15–18 hectolitres per hectare. The vineyards are on the schist of the Latour-de-France. Brigitte explained that they work organically, and were certified in 2010, but only for their vins secs. If your Vin Doux Naturel is organic, the alcohol that you use to stop the fermentation must also be organic, and organic alcohol is so much more expensive than ordinary alcohol. They simply cannot afford it, or justify the significantly higher price to their customers.

First, I tasted the vins secs. The white wine, Ver-tige, Côtes Catalanes, is Macabeo blended with some Grenache Gris to give structure. They make three single varietals for red wine, Grenache Noir, Mourvèdre and Carignan, and are also planning a Tannat, the grape variety of Madiran, which is also allowed in Côtes Catalanes. There are also blends, a spicy Caramany and a more structured Latour-de-France.

However, you sense that Brigitte's real love is Vin Doux Naturel. First, we talked about *rancio.* Rancio Sec has been a permitted category for Côtes Catalanes since 2004. Unlike Rivesaltes Rancio, it is not muted, but oxidized, and aged in a vat for ten years. Brigitte's classic wine is a Macabeo Côtes Catalanes Rancio Sec. You could compare it to old sherry, with firm acidity, dry, nutty fruit and an incisive note on the finish. A more unusual Carignan Rancio, which had spent ten years in barrel, was a deep reddish brown, with a sweet nose and more nutty fruit.

The first Vin Doux Naturel was Le Temps d'un Oubli, a muted Macabeo that had spent five years ageing in barrel. It was rounded and supple with notes of orange and apricots, and less incisive than the non-muted Rancio. For the Vin Doux Naturel a long élevage is critical. Their youngest wine is four years old, and their oldest wine currently in bottle is from 1948. It is fabulous, with concentrated flavour, rich and long with balancing acidity. Brigitte is sad that people do not believe in Vin Doux Naturel – even the people who make the wines are not proud of them. She put a lot of the problem down to the fact that people do not know when to drink Vin Doux Naturel; in fact, it can accompany a lot of different dishes and flavours. She lamented how difficult they are to sell. But where else would you find a 70-year-old wine for €200 and be sure that it was still going to be a highly enjoyable drink? At times, it is certainly an uphill struggle.

As I left, I spotted a notice in the tasting room – *il vaut mieux boire du vin d'ici que de l'eau de là.* It is better to drink the wine from here, than the water from over there.

Roc des Anges

Latour-de-France

www.marjorie-stephane-gallet.com

Marjorie and Stéphane Gallet created their estate in 2001. She comes from the northern Rhône, and met Stéphane while studying at Montpellier. After graduation he went to work at Mas Amiel, so looking for vines in Roussillon was the obvious decision. Côte Rôtie would have been much more expensive, and as they note 'here in Roussillon everything is possible … the schist is the same as in Côte Rôtie, but the climate is different.'

Since 2014 they have worked in a smart cellar near the village of Latour-de-France, with lots of concrete vats of various sizes to accommodate their numerous different parcels of vines. Altogether they have about 100 plots in just 35 hectares, at altitudes between 150 and 350 metres; 60 ares is a big plot for them. Virtually all their vines are in Montner, with a few in Tautavel and also close to Calce. Marjorie was at pains to point out that the somewhat unkempt vines next to their cellar are not theirs. Montner has very poor schist soil; and yields are low. They also have a vineyard in Tautavel, Terres de Fagayra, for Maury.

Marjorie and Stéphane have Grenache Gris and Macabeo which have been replanted by massal selection, as well as Carignan Gris, Carignan Blanc and Grenache Blanc. For reds there is Carignan Noir, which they have also replanted, and also Grenache Noir and Syrah, as well as a little Cinsault, which they planted in 2015. Cinsault is unusual in Roussillon as it is not suitable for Vin Doux Naturel, but Marjorie thinks that it may give them the refinement they are seeking: 'The soil here gives elegance,' she observed. She admitted that they would not replant Syrah, stating that 'the truly Mediterranean varieties work best.' Ninety per cent of their vineyards are on grey schist, with a little limestone, as well as granite and some *galets roulés*. Their highest vines are at 300 metres, north facing and closer to the mountains than to the sea, so that they enjoy the effect of cool air, but no maritime influence. The average vine age is 65, but a lot have been replanted. They have been certified organic since 2010, and have worked biodynamically since 2011.

For the red wines, they try to leave the berries as intact as possible, while removing the stems, so they use an *égrenoir,* which just drains, as opposed to an *égrappoir*. Carbonic maceration starts inside the berries, making for earlier drinking wines. This is something they have done since 2008. Marjorie does not like eggs as the lees move, which she believes tires the wine; she likes to keep her wine on still lees during fermentation, whereas in eggs the wine gets 'too fat and loses its energy'. She also dislikes amphorae as they soften the tannins. She mainly uses barrels for white wine, and is increasingly doing a malolactic fermentation. She admitted to being scared of it initially, thinking it would lower the acidity, but she has found that biodynamic viticulture helps retain acidity levels. There is an extensive underground barrel cellar at Roc des Anges. Marjorie is adamant that where a wine is aged has a huge impact on its flavour; they checked the ley lines when building the cellar, and found it to be in a particularly favourable spot.

Llum, meaning 'light' in Catalan, is a blend of Grenache Blanc, Carignan Blanc and Macabeo, all picked at the same time, with fresh elegance and mineral tension. Pi Vell, meaning 'old pine tree', comes from Macabeo planted in 1947; in 2017, 70 ares produced just 6 hectolitres. L'Oca, 'goose', from a vineyard shaped like that bird, is also from Macabeo, but planted in 1957 on red clay as well as schist. It was firm and stony, but with less tension. Marjorie really likes Macabeo; some people criticize it for its neutral flavours, but for her it really expresses the soil differences.

Segna de Cor, a blend of Grenache, Carignan and Syrah from relatively young vines, aged in tank, had some dry spice, with elegant structure. The name is an anagram of Roc des Anges. Reliefs is mainly Carignan, from vines aged between 40 and 100 years, with firm, stony fruit and a tannic streak. The wine called 1903, from three different plots of Carignan planted in that year, is rounded with red fruit. Marjorie favours a relatively short maceration of about a week, and as the maceration is short, adds the pressed juice, believing it makes the wine complete.

In Maury, they concentrate on Vin Doux Naturel, with a rich but elegant Grenat. Marjorie thinks that biodynamics make the sugar lighter, so that there is no alcoholic burn. Opus Nord, from Grenache Noir with some Carignan, all planted together, has deeper fruit. Maury Blanc from Grenache Gris, Grenache Blanc and Macabeo is lightly biscuity. For Marjorie, Maury is not a suitable area for vin sec, as you cannot pick the Grenache early enough. Our tasting finished with a 1996 Maury Tuilé, aged in cement vats and foudres without any *ouillage* and bottled in 2016, which had some dry nutty fruit and a harmonious finish. They also make a Maury Rancio Sec, from Grenache Gris and Macabeo, which spends a minimum of five years outside. *Flor* is not essential and the wine is bottled when it is sufficiently salty and intense.

Clos du Rouge Gorge

Latour-de-France
cyrilfhal@gmail.com

Cyril Fhal has his cellar in the tiny hamlet of Cuxous, on the road between Cassagnes and Latour-de-France. You can make out the rather forbidding walls of the medieval château of Cuxous through the trees. There was a wonderful atmosphere in his cellar, created by the voice of a nun singing ecclesiastical chants. It is not for the wine, Cyril explained, but for him, and his cellar workers. The music varies; his taste is quite eclectic.

You immediately sense that Cyril is a thoughtful, sensitive winemaker. 'It is a lie to say that I am in control; we juggle; every year is different,' he says. He related how he came to Roussillon. The name Fhal is Alsacien in origin, but his family were in Algeria and returned to Paris before settling in the Limousin. Cyril was trying to decide what to do, and in 1995 worked a vintage in Saumur, which really fired his interest, leading him to study in Bordeaux. After that he did a *stage*, two years alternating work and study at Vignobles Laffourcade in the Loire Valley. When the *stage* finished, they asked him to stay on and run the three estates. His neighbour was Marc Angeli, one of the leading lights of the natural wine movement. Another job in the Loire followed and while he was wondering what to do next, he came to explore the hills of Corbières, which led him first to Perpignan and then to Calce, where he was welcomed by Lionel Gauby. He fell in love with the landscape and took a job as *régisseur* at Mas Crémat. He found 1.5 hectares of vines of his own and a house in Latour-de-France with a tiny cellar.

The growth of his estate has been a gentle progression. The cheapest vines are the old vines on slopes, which give very small yields and are difficult to work. Cyril is keen to save old vines and has now amassed 8 hectares altogether. It can be worrying, working with them, he says; they are 'the geriatrics of viticulture'. Some he has pulled up to replant in ground on which he has grown cereals and legumes for six years. Le Col du Loup is a plot of Grenache Noir planted by massal selection, a clos with high walls that is home to lots of robins, giving the estate its name. For red wine, Cyril has Grenache, Carignan and a little Cinsault, and for white wine, Macabeo, the most important variety of Latour-de-France, as well as Grenache Gris, Grenache Blanc and Carignan Blanc. Discussing the amount of work needed in the vineyards Cyril noted that here five people look after 8 hectares, but in a mechanized vineyard, you only need one person for 15 hectares. Such are the demands of elderly vines.

Most years Cyril makes two white wines and three reds. The first white wine is a blend of Macabeo with some Carignan Blanc from 80- or 90-year-old vines. Cyril wants tension in this wine; he picks at the beginning of the ripening period, at a potential alcohol of 12–12.5% abv. The picking date is crucial. The 2018 did indeed have tension and elegance, and the restraint of Macabeo, and will develop with age. The fermentation begins in vat, without any *débourbage* and then the wine is aged in small foudres or barriques until bottling in June. A good barrel adds length and weight, allowing a wine to develop, says Cyril. He made the analogy of a tailor bringing out the best in a body, and said that the cooper does the same for a wine. L'Ubac is the *grand cru* of the estate, a north-facing slope with 150 metres difference in altitude; the red vines are at the bottom and the white at the top. The wind can have a drying effect here, quickly changing the balance in the grapes. White L'Ubac is Grenache Gris and was richer and more intense, with exotic flavours and great length.

Jeunes Vignes was very perfumed and ripe, with fresh acidity and tannin, from Grenache picked at a potential alcohol of 13–13.5% abv. Cyril ferments whole bunches, with a gentle extraction, a semi-carbonic maceration in foudres and some *pigeage*: like tea, he said. The maceration takes about ten days and wine goes into barrel while some sugar remains, which adds texture. His Cuvée Classique comes from 100-year-old Carignan vines, which needed a lot of work. Carignan is a difficult variety, but not necessarily a great one – he called it *une casse-tête chinoise*, or a Chinese puzzle that challenges the wine grower. Red L'Ubac, of which he makes just 600 bottles, was elegant and structured, coming from a 115-year-old plot on the aforementioned slope, with a bit of everything; Carignan, Macabeo and Clairette but mainly Cinsault. Sadly, the time constraint of our next appointment did not permit a visit to the vineyards; I hope to have that pleasure next time.

Domaine Jean-Louis Tribouley

Latour-de-France

04 68 29 03 86

Jean-Louis Tribouley opened his anonymous front door, with a number but no name plate, and we followed him down a narrow corridor and out into a courtyard. Here we were greeted by a tortoise named Elvis, who was very taken with my pink toenail varnish. Off the courtyard we entered Jean-Louis's small, cool cellar.

Known as a natural winemaker, Jean-Louis is one of the best among a group of about 18 such winemakers in and around Latour-de-France. He referred to them as a network, or a mafia, depending on your point of view. For him natural winemaking entails organic viticulture, natural yeast, impeccable hygiene and the use of sulphur only when absolutely necessary. Winemaking represented a career change for Jean-Louis. He had done lots of other things, including working as a carpenter and for the social services before it was time to change direction again. He had no experience of winemaking, but simply liked wine, so he went to Beaune for a diploma in viticulture and oenology, came south to help Olivier Pithon prune in January 2001 and then spent six months with Gérard Gauby. Originally from Besançon, Jean-Louis liked the village of Latour-de-France and found 11 hectares of vines there and in the neighbouring villages of Cassagnes, Rasiguères, Montner and Estagel. Altogether he has 17 different plots: two plots make up 6 hectares, and 15 plots account for the other five. He made his first wine in 2001. He talked about the geological crossroads, with gneiss in the higher vineyards, schists in lower vineyards, as well as clay and limestone. His vineyards are at altitudes of between 50 and 250 metres.

Above all Jean-Louis enjoys single varieties: 'You can find the truth in a *monocépage*,' he says. He has no time for the appellations, or even the IGP Côtes Catalanes – 'they make for standardization and are too complicated' – so all his wines are Vin de France. 'The official commissions don't understand that people are making *vins de terroirs*,' he grumbles.

Our tasting began with La Part de l'Eté, a pét nat made from Syrah, Grenache and Carignan, which was light pink in colour with refreshing raspberry fruit and a lightly creamy finish. The method is the same as the ancestral method of Limoux; the fermentation finishes in the bottle, followed by *remuage* and disgorgement to remove the lees. His white wine, GG for Grenache Gris, with 20 per cent Macabeo, had firm salinity and an underlying richness, after a fermentation in old barrels. The colourfully labelled Les Copines is pure Grenache Noir, with a semi-carbonic maceration, in other words an eight-day infusion, making for rounded, ripe cherries and some dry, peppery spice. Elepolypossum, a pure Carignan, was ripe and rounded, but not yet very harmonious, and Jean-Louis observed that 2019 was not such an immediate year as some. And why the name? Jean-Louis explained that a friend, an Italian artist, had designed the label for him, with its very curious animal, representing an image of Carignan, after an evening of many bottles. His children suggested an elephant, an octopus or polyp, or a possum, and Elepolypossum was the result, with another friend suggesting that could mean *ensemble on peut plus*, or together we can do more.

Essentially, Jean-Louis said, there are very few decisions to be made. He wants healthy vines and healthy grapes – 80 per cent of the wines' quality comes from the vineyard, 10 per cent is the work of the vigneron and the final 10 per cent he attributes to *le hazard*, luck and circumstances, including the intuition of the vigneron. 'You can't lie with wine; there is a magic moment,' he says.

TARERACH

Domaine Ronald Joachin

Tarerach

joachinronald@yahoo.fr

A chance bottle at dinner with a winemaking friend in Faugères on a Saturday evening led us to a meeting with Ronald Joachin in Latour-de-France on the following Tuesday. We had greatly enjoyed the bottle, intriguingly labelled C'est un cépage blanc catalan, guessing the grape variety in question to be Macabeo. Happily Ronald's email address was on his label. As he admitted when we met, he couldn't afford business cards, so very sensibly put his contact details on his labels.

He comes from Arles, where he still lives, and used to work in the restaurant trade, which led to a desire to make wine. Various *stages* followed, with Catherine Bernard near Sommières, at Terres Promises in the Var and with Dominique Hauvette in Les Alpilles, whose six hectares of vineyards in Roussillon he subsequently bought, in time for a first vintage in 2018. For his cellar, he has taken over part of the old cooperative building in the village of Tarerach. As well as Macabeo he has Lledoner Pelut, Syrah and a little Carignan.

The Macabeo is kept in vat and has firm, stony acidity. Ronald explained that it is the result of three pickings, at potential alcohol levels of 11.5%, 12.5% and 13.5% abv, three vinifications and a final blending. There was some firm minerality and youthful structure. L'Arsouille, a description of someone who is overfond of the bottle, is a blend of Lledoner Pelut and Carignan, with fresh red fruit. C'est un cépage noir catalan comes from Lledoner Pelut with some spicy fruit and a touch of tannin. Ceci est un cépage nord de la vallée du Rhône is a pure Syrah, with fresh, peppery fruit. Ronald believes in minimum intervention. In

2018 his red grapes were destemmed, but not in 2019, and he is experimenting with an orange Macabeo. It had already been on the skins for nine months but was still light golden in colour – the deep colour of orange wines comes from the oxidation – this had tannin, acidity and structure with some firm fruit, and promised well.

You sense that Ronald has lots of ideas. He is thinking about a sparkling Macabeo – 'Does anyone else make one?', he mused – and also considering ageing Macabeo in barrel. He has certainly not chosen an easy path, but I do hope he does well. From the quality of his wine, he certainly deserves to.

Mas Llossanes

Marquixanes

www.masllossanes.fr

Mas Llossanes was my very last discovery for this book. Dominique Génot contacted me via the website of Millésime Bio in January 2021 and we talked on Zoom. He remembered that we had met once before at a large tasting of Italian wines organized by *Decanter* magazine, when he was running the estate of Caiarossa at Riparbella near Pisa. Dominique comes from Lorraine, but after 11 years enjoying the Mediterranean climate of Tuscany, the prospect of returning to the north of France was impossible, so he and his wife Solenn looked for vineyards in the south. In Roussillon it was the proximity of both the mountains and the sea that really appealed, and the vineyards were affordable, so in 2016 they made their first wines from 11 hectares of vines outside the village of Tarerach. Their vineyards are some of the highest of Roussillon, lying at altitudes between 550 and 700 metres, in one large plot with two soil types, gneiss and schist. The vines are relatively old, between 25 and 80 years. Over a third of their vineyard is planted with Carignan, including some Carignan Gris, and they also have Grenache Noir and Lledoner Pelut, with a few vines of Grenache Gris as well as old plantings of Syrah and Cinsault. There is a little Chenanson, which is a cross between Jurançon Noir and Grenache Noir, and for white wine, just one hectare of Chasan, a cross of Chardonnay and Listan.

The vineyard is farmed biodynamically and in the cellar Dominique uses natural yeast and as little sulphur as possible. He makes one wine, Là-Haut, a pure Carignan destined for early drinking, without any sulphur at all. For his white wine, the Chasan is partly fermented and aged in oak; his rosé is a pure Cinsault, and there is a range of five red wines, including Là-Haut. Au Dolmen – there is a 5,000-year-old Neolithic dolmen in the middle of his vineyard – is a blend of predominantly Carignan, with some Syrah, Chenanson and Grenache, with an élevage mainly in vat. Dotrera is a similar blend, but aged mainly in barrel, and Le Pur Carignan and Le Pur Syrah are just that, pure expressions of those two varieties, made only in the best years, and in tiny quantities.

Dominique enthused about Carignan. He notices considerable differences within his vineyards, so that he makes Là-Haut from the more vigorous and higher yielding vines, while Le Pur Carignan comes from less productive vines, with more concentrated flavours. There may be two or three pickings of Carignan during the harvest, with as much as a week between them. Generally Dominique favours a classic vinification with a three-week maceration, making for a slow extraction with very few *pigeages* or *remontages*. Most of the berries are lightly crushed, but some intact berries, allowing for some intercellular fermentation as in a carbonic maceration, make for different aromas and add complexity. Most of his wines are Côtes Catalanes, rather than Côtes du Roussillon and his labels depict a pair of pruning secateurs, making a dramatic image.

Dominique has just taken over the old cooperative building in the village of Marquixanes in the Têt Valley. The first thing he has to do is remove some of the enormous old concrete vats. I am looking forward to a cellar visit when we are finally able to travel again.

CASSAGNES

After a detour to Tarerach it is a short, scenic drive to Cassagnes from Latour-de-France, and then on to Bélesta.

Domaine Modat

Cassagnes

www.domaine-modat.com

Philippe Modat is a Catalan who returned to his roots. His grandfather had vines and Philippe's legal studies in Bordeaux triggered his own interest in wine. In partnership with his father he bought land and vines in 2007 from a cooperative member in Cassagnes. The following year he built a streamlined cellar outside the village with stone walls and solar panels and surrounded by eight hectares of vines. After dividing his time between wine and law, Cassagnes and Paris, Philippe has returned to Paris and his two sons, Quentin and Louis, have run the estate since

2015. Louis, an agricultural engineer, studied at Toulouse and did various *stages*, including Château Palmer, Domaine de Beaucastel and in New Zealand. Quentin's strength is marketing and administration and you immediately sense a good rapport between the two brothers.

They now have 30 hectares of vines, all of which are around Cassagnes except for their grandfather's original three-hectare vineyard at Polestres in Les Aspres. The vineyards have been farmed biodynamically since 2012. Cassagnes is part of Côtes du Roussillon Villages Caramany, where gneiss is the main soil, with a hard bedrock and vineyards at about 300–400 metres altitude. Historically Caramany is dominated by Carignan, with less Grenache. The brothers are also developing their white wines, but do not make any Vin Doux Naturel. However, Rancio Sec interests them.

We tasted with Quentin in their barrel cellar, which has 600-litre demi-muids as well as foudres and small barriques. Quentin is an articulate exponent of his wines. First, we tasted De-ci De-là, a white Côtes du Roussillon, made from the traditional varieties of the region, mainly Grenache Gris and Grenache Blanc, but also Macabeo and Carignan; only Tourbat is missing but they are planning to plant it. It was a classic white Roussillon, with firm mineral notes, acidity and texture. Next came Les Lucioles, from two young vineyards of Viognier and Roussanne that were planted in 2008 with cuttings taken from Condrieu and Hermitage, along with 10 per cent each of Macabeo, Carignan and Grenache. It was rounded and peachy with fruit and texture. Quentin observed that although Viognier and Roussanne are allowed in the appellation, they do not consider them typical so prefer to label the wine Côtes Catalanes.

A rosé followed, Le Petit Modat Amour, or *mot d'amour*, from Grenache and Syrah aged in old barrels. They want their rosé to taste like a white wine, with freshness and weight. Grenache works well for rosé but Carignan is less successful. Le Petit Modat Amour Blanc is based on Muscat, with some Viognier, Roussanne and a little Grenache Blanc. It had some grapey fruit and a rounded palate; Quentin called it a modern version of Muscat. In 2019 they made just one barrel of a firm, mineral new white wine, Nouvelle Cuvée No 1, from Carignan Blanc. Nouvelle Cuvée No 2 – the names have not yet been decided – is Grenache Noir vinified as a white wine, with some Grenache Blanc and Grenache Gris. It is fresh and rounded. Quentin told me they favour natural yeast in their winemaking, with no filtration or fining, and use as little sulphur as possible.

Then we went on to red wines, with Le Petit Modat Amour, a blend of Grenache, Carignan and Syrah, kept in a stainless steel tank. It was rich and round, and eminently drinkable; Quentin called it the DNA of Roussillon. They produce two-thirds red to one-third white. Comme Avant is a blend of Carignan from vines that are between 80 and 115 years old, blended with some Grenache and Syrah, and kept mainly in concrete vats, or old wood, for 18 months. It had intense black fruit, with a fresh finish and a youthful balance. Sans Plus Attendre is based on Syrah with 15 per cent each of Carignan and Grenache Noir to produce deep colour, intense black fruit and peppery flavours.

A pure Carignan, Les Grognards, meaning *ce qui grognent*, or those who growl, a reference to the veterans of Napoleon's army, had elegant fresh fruit. It comes from the plot of vines by the cellar, which featured on the 1905 *cadastre*, so they know that the youngest it could be is 115 years old. Le Plus Joli is literally that, a selection of the three best barrels. For 2016 that was two Syrah and one Carignan, but the choice varies from year to year. It was firm, youthful and intense.

Quentin talked about their projects and experiments; you sense they have lots of ideas as well as youthful enthusiasm. He mentioned a pét nat, a Rancio, a solera. I left feeling that the change of generation had been successfully effected and that the estate was in very good hands for the future.

BELESTA

Clot de l'Oum

Bélesta

www.clotdeloum.com

Eric Monné's cellar and house are on a slope outside the village of Bélesta and from his living room, you have Roussillon at your feet, with views of Força Réal and the Pyrenees. The view during our tasting was simply breathtaking.

Eric and his Brazilian wife, Leia, bought their vineyards in 1995 and initially sold their grapes to the cooperative in Cassagnes, before creating their estate in 2001. They have 12 hectares in production, all within a two-kilometre radius, and planted mainly on granite, at altitudes between 350 and 600 metres. They have been farmed organically for 20 years, which improves the pH levels in the wine. Eric observed that the soil of the Fenouillèdes hills is very mixed; they were formed at the same time as the Pyrenees. Clot de l'Oum means the valley of the Orme – in this instance a *clot* is a valley, not a *clos* or enclosed vineyard, and *orme* is the French for elm. Eric once worked for the patent office in The Hague but, coming from Perpignan, he wanted to do something in Roussillon, so why not wine? He was not deterred by the fact that he had never studied oenology.

Eric has helped create the reputation of white wine in the Fenouillèdes. First came Dolce Gavatx. Eric laughingly explained that *gavatx* means a foreigner, a stranger, in Catalan and that people from Narbonne are considered *gavatx* in Perpignan. The wine is a blend of Macabeo and, unusually, Carignan Gris, which Eric planted using massal selection. The wine is very fresh and mineral, fermented in vat and finished in old barriques. Eric is convinced that the reputation of Roussillon will be, or perhaps even is being created by its white wines, with the indigenous varieties Macabeo, Grenache Gris and Carignan Gris benefiting from the altitude and the influence of the Pyrenees. Cine Panettone is mainly Carignan and Grenache Gris, with more depth and structure than Dolce Gavatx.

The 2017 Gavatx Rouge, a Côtes du Roussillon Caramany, comes mainly from Carignan, with a classic vinification of destemmed grapes and a short maceration making for fresh fruit and a streak of tannin. In comparison the 2018 Gavatx is based on Grenache, again with some elegant fruit. Eric looks for freshness and good acidity in his wines, both red and white, observing that if you work the soil, you can pick earlier.

La Compagnie des Papillons, named after a French children's game, is a Côtes du Roussillon Villages Caramany, mainly from Carignan and Grenache and aged for 12 months, making for some ripe fruit and spice. Then, to show how well the red wines of Roussillon age, Eric opened the 2010 Numéro Uno, a blend of 85 per cent Syrah with some Carignan, that had spent 15 months in barrel. The nose and palate were beginning to develop attractive notes of maturity with some leathery, peppery flavours. St Bart comes from 100-year-old Carignan, with some Syrah, aged in barrel for elegant fruit and a fresh finish. Numéro Uno has been replaced by Le Clos; mainly from Syrah, it is aged in foudres rather than barriques, producing some firm structure and ageing potential.

I quizzed Eric about the changes in his winemaking and he talked of looking for elegance. There is no problem with obtaining tannin or colour, but freshness can be more of a challenge. I had no doubt that he is succeeding.

Domaine Riberach

Bélesta

www.riberach.com

Riberach was the old name of the village of Bélesta, which I approached on this occasion from the direction of Ille-sur-Têt, enjoying a dramatic view of the hilltop village. It was once a frontier town between France, or Occitanie, and Catalonia. In the 1970s it had 600 hectares, half of which produced Vin Doux Naturel.

Domaine Riberach was set up by four associates in 2006. Jean-Michel Mailloles is a wine grower in the village of Bélesta, with his German wife, Karine, who has financed the investment. An Austrian sommelier, Mauritz Herzog, was involved initially, but he has returned to Austria. The wine is made by Patrick Rodrigues and he has been recently joined by Guilhem Soulignac, with whom I spent a happy couple of hours chatting and tasting. Jean-Michel continues to run the vineyards. They have taken over the old cooperative cellars of the village, and in addition to a spacious cellar the building houses a smart hotel with bedrooms created out of the indestructible old concrete vats, and one of the better restaurants of Roussillon. The building dates back to the 1920s and 1930s, a time when every family in the village had vineyards and also goats.

Guilhem talked about what he called the *côté mosaique*. In their ten hectares, they have three or four different terroirs, brown schist around Montner and Latour-de-France, granite and limestone in Calce and also some gneiss. Some of their wines come from specific vineyards and some are blends. The range has evolved. In 2006, they made only red wine and then followed with an experimental white wine in 2007. These days they produce an equal quantity of red and white wine, with about 10 per cent of production being rosé. For white wine they have mainly Macabeo, Grenache Gris and Grenache Blanc, as well as Carignan Blanc and Carignan Gris, of which they have planted more. For the red wines they work with Syrah, Grenache and Carignan.

I tasted with Guilhem in the rather smart shop attached to the hotel. All their wines, except for a sparkling Vin de France, are Côtes Catalanes as they do not meet the appellation requirements. Synthèse Blanc is a blend of Grenache Blanc and Grenache Gris, with some Macabeo and a little Carignan Gris, grown on either schist or granite. The wine undergoes a malolactic fermentation and 70–80 per cent of it is aged in barrels of various sizes, tronconic vats, and 500- or 700-litre demi-muids. The wine is very intriguing, with the bitterness of the Macabeo and some salinity and fresh acidity on the finish. Hypothèse Blanc is a pure Carignan Gris, of which 40 per cent of the blend comes from 100-year-old vines grown on schist, while the rest is from young vines on granite. It is kept in 12-hectolitre foudres and a few demi-muids and the resulting wine is quite severe with a pronounced freshness, some oak and some acidity.

Rouge No 18 is mainly Syrah grown on schist, with some Grenache, also from schist, as well as young Carignan on granite. The Syrah is kept in vat and the Carignan and Grenache in demi-muids, making for a rounded, spicy wine with red fruit and a fresh finish. Antithesis comes from Syrah grown on limestone in two plots between Bélesta and Montalba in a valley sheltered from the tramontane wind. It has intense black fruit and a fresh, peppery finish. Synthèse, aged mostly in barrel, is based on old Carignan grown on granite, with some Syrah grown on schist, and a little Grenache, also grown on schist. Red fruit flavours are balanced with elegant concentration and freshness from the Carignan. Hypothèse Rouge comes from a plot of Carignan planted in the middle of the nineteenth century by Jean-Michel's great-grandfather. The soil is granite, the average yield is 15 hectolitres per hectare and the wine is aged in 600-litre demi-muids, making for rounded fresh fruit and balanced tannins. Parenthèse is an orange wine, from Grenache Blanc, Macabeo and Grenache Gris, from destemmed grapes given three weeks maceration. It was fresh and dry with a fragrant note on the finish.

Guilhem observed how their winemaking has changed, with less extraction, shorter macerations, less sulphur, more careful élevage and no more small barrels. They are working towards more site selection as they get to know their vines better.

Guilhem makes his own wine, Prolegomenes, from grapes that he buys from Jean-Michel. It is a blend of Grenache Gris and Macabeo with some Carignan Gris, grown on granite and given some skin contact. It was very linear and very direct, and a lovely example of how white wine is developing in Roussillon, as were all the white wines of Riberach.

MONTNER

Le Débit d'Ivresse

Montner

www.ledebitdivresse.com

A chance bottle at our local wine bar, Le Picamandil, in the village of Puissalicon led to a meeting with Luc Devot. The wine was called

Hydrophobia and the label was a take on Munch's painting *The Scream*. It was a delicious blend of Chardonnay and Grenache Gris, with some fresh fruit and incisive acidity. It tasted just as good in Luc's cellar.

Luc makes his wine in part of the old cooperative building in Montner. He has 4.5 hectares and also some buys some grapes, producing a characterful range of wines. And how did he come to wine? On his own website, he talks of meeting Jean-Louis Tribouley, who introduced him to the concept of natural wine, before working at Domaine la Petite Baigneuse in Maury and making his first wine in 2017. He had been a cook, fisherman, musician, animator, educator, before becoming a winemaker. The name of his estate translates as the flow of drunkenness.

Amb Amics, meaning *avec des amis*, or with friends, in Catalan, is a blend of Syrah with 15 per cent Grenache. It was light red in colour, with fresh fruit and refreshing acidity, after a seven day carbonic maceration. Goulou Goulou, with the name recalling the cartoon characters Glouglou, is a pure Syrah with some fresh fruit. My Favourite Things, a blend of Grenache with 20 per cent Mourvèdre, was quite firm and tannic, balanced by some red fruit. The final wine was L'Appât du Grain, which could translate as 'the lure of the grape', a blend of Carignan with 30 per cent Mourvèdre, with some firm fruit. You certainly sense Luc's creativity with his wine labels, which translates into his evident enjoyment of his current career. We left him heading for a meeting to plan the next Portes Ouvertes in Latour-de-France.

Domaine de l'Horizon

Montner

www.domaine-horizon.com

Thomas Teibert had left his cellar keys at home, a half hour drive away, and for a moment it looked as though our tasting might not happen. Happily, the *mairie* down the street lent him a ladder, enabling him to climb over the wall, and we were able to taste after all. Thomas comes from Ulm, near Stuttgart, which is a region of beer drinkers, but he saw a film called *The Winemaker* on the television at an impressionable age of 12 or 13, which led to a visit to VinExpo in Bordeaux when he was 16. Oenology studies followed, and he worked in the Alto Adige for a leading producer, Manincor. Then he met Gérard Gauby and came to Calce, where he fell in love with the scenery, and also with Gérard's daughter, Mélanie. He has never left, with 2007 his first vintage in Roussillon.

The name Domaine de l'Horizon was suggested by Ghislaine Gauby, Gérard's wife. An old wine grower in Calce used to refer to the views as *mon horizon*, and Thomas can see the sea from nearly all his vineyards. He now has 17 hectares of vines, 14 in Calce and 3 at the far end of the Agly Valley in Caudiès, with no plans for further growth. He has just moved into Tom Lubbe's old cellar in Montner. You immediately sense that Thomas is a thoughtful winemaker. He wants elegance and precision, nothing either too natural or too oaky. He looks for finesse and a sense of place, of terroir. He observed that it is a fashion in Roussillon to pick the grapes before they are ripe; Thomas is restrained, there is a two or three-day window, so you should pick at the beginning of the window, he explains. If the grapes are all at the same degree of ripeness, the wine will be monotonous. He destems and uses indigenous yeast. He is not in favour of orange wine, as it leads to uniformity of flavour.

As well as making wine, Thomas is also the agent for the Austrian coopers Stockinger, which gives him firm ideas about the use of barrels, and how they are made. It is a craft, he says; you must cut the trees *à la lune*, with the full moons of Christmas and February when the wood is at maximum density. Stockinger do not use artificial driers so there is a long drying period; oak is a powerful anti-oxidant. The relationship between oak and wine must be managed well but when it works it is very interesting. Naturally Thomas uses Stockinger barrels for his own wines.

All his wines are Côtes Catalanes: he does not think that the appellation system works well, criticizing it for insisting on the inclusion of Syrah in the appellations, as though Grenache and Carignan are worthless. His view is that the appellations should defend authenticity. Our tasting began with L'Esprit d'Horizon Blanc, a blend of 80 per cent Macabeo with Muscat à Petits Grains. Macabeo is very neutral while the Muscat adds some fruit and a note of honey. The wine was nicely characterful, tight-knit, with firm mineral notes and good acidity. A second white, from a field blend of Macabeo, Grenache Gris and Grenache Blanc, as well as some Carignan Blanc, spends 12 months on the lees in tonneaux, demi-muids or foudres. It was rich and structured, while a Réserve Personnelle, from a similar blend, but receiving a longer élevage of four years in foudres plus a year's bottle age, was rounded and solid, with good length. Thomas began by using foudres, but now much prefers smaller *pièces*. Mar i Muntaya, the sea and the mountain, with a label drawn by his 6-year-old daughter, comes from vines grown in Caudiès, Syrah with 10 per cent Grenache, using whole bunches and a semi-carbonic method of vinification, with no *foulage*. Thomas called it his Beaujolais: it was fresh, with cherry fruit and some acidity, and bottled in the spring after the harvest. Esprit de l'Horizon Rouge is a pure Syrah, with fresh, peppery fruit. Thomas commented that he did not like Syrah from the south and this certainly tasted more of the north, with some fresh fruit. In 2018 he made his first wine without sulphur; it spent 8 months in a 15-hectolitre Stockinger foudre.

The 2014 Horizon Rouge, from two-thirds Carignan vines aged between 70 and 100 years, and one-third Grenache vines that are 60 years old was aged for 12 months in demi-muids and foudres. It had a fresh, structured palate with appealing red fruit. The 2016 vintage was more intense, youthful and dense. Thomas considers Carignan to be complicated, and even more of a diva than Pinot Noir. He is adamant that a winemaker is a craftsman, combined with an artist, who must interpret the terroir. The terroir of Roussillon is very heterogeneous, and there are many different styles and estates. Thomas is a thinker, a philosopher, and gave me food for thought as I drove to my next appointment.

Domaine Latri Fontoj

Montner

latrifontoj@gmail.com

Stijn Demeulemeester, with his German wife Franziska, is one of the newest arrivals in Roussillon. A chance meeting *chez* Jean-Louis Tribouley in Latour-de-France led to our cellar visit in Montner. Stijn comes from Ghent in Belgium, and has worked in both France and Belgium as a sommelier but really wanted to make his own wine. And why Roussillon? He had met Frédéric Rivaton, another wine grower in Latour-de-France, with whom he learnt to prune. Like so many before him, he fell in love with the region. Land is cheap, but what was not so easy was finding a house with a cellar.

Latri Fontoj means *trois sources* or three springs in Esperanto and they have three hectares in four main plots. There is Muscat à Petits Grains just outside Montner, a *complanté* vineyard of Carignan, Grenache Gris and Grenache Blanc, 40 ares of Syrah on the slopes of Força Réal at 300 metres, and finally some Mourvèdre on a north-facing slope at Rasiguères. Stijn made his very first wine in Roussillon in 2017, while doing a *stage* with Gérard Gauby. Previously he had worked in Chile, making Pinot Noir and Pais. His wines are Vin de France, to keep life simple. His first commercial vintage was 2019, which is what we tasted.

Pepe is a rosé from Grenache Noir, with a bright colour and fresh red fruit and acidity, rounded out by three months in barrel. Stijn favours minimum intervention, natural yeast and no fining, filtering or sulphur. Bamm Bamm, with a reference to *The Flintstones*, is also pure Grenache, picked later than for the rosé, with some fresh cherry fruit. The maceration was two to three weeks, it was followed by a gentle infusion and then the wine spent six months in barrel.

2018 Tulpe, a pure Syrah, was more solid on the palate and nose, with some red fruit, after a two-week maceration and 12 months in wood. Stijn generously opened his very first wine, 2017 Bonne Pioche, a blend of Grenache and Mourvèdre. The fermentation was layered, with whole bunches and crushed grapes, and then the wine spent 12 months in barrel, making for ripe rounded fruit with a good balance of acidity and tannin. Stijn had new plots for the 2020 harvest, the Muscat and the Carignan; he is also interested in varieties like Lledoner Pelut and mentioned Trepat, a grape variety from southern Catalonia. I thought he had made a good start.

CALCE

The village of Calce is not on the way to anywhere. You really have to want to go there but there are plenty of reasons to visit for it has become a focal point in the development of vin sec in Roussillon, thanks above all to one pioneering wine grower, Gérard Gauby. He has acted, quite unintentionally, as a magnet for other wine growers and at the last count there were five wine cellars in the village, as well as the small co-operative and several other notable wine growers with vineyards in the village. The lively restaurant, Le Presbytère, in the centre of Calce is also worth the detour. The scenic route from Estagel takes a narrow, winding road, with barely room for two cars to pass for most of the way; I held my breath and hoped that I would not meet a tractor coming in the other direction.

Face B

Calce

www.vins-face-b.fr

Séverin Barioz was born in Lyon, and trained as a lawyer, specializing in viticultural law. He worked in Paris and then in Burgundy for a number of years as the director of the Syndicat des Vignerons de la

Bourgogne, before changing directions to become a winemaker or, as he put it, switched sides, 'passer de l'autre côté'. He has a friend in common with Jean-Philippe Padié and first helped Jean-Philippe with the harvest in 2004, and returned every year. He fell in love with the village, feeling it to be quite special and so took the decision to stay. He made his first wine in 2015 from just four hectares of vines, in six different plots between Calce and Baixas.

Why Face B? It's the B-side of a vinyl record, as well as what Séverin called the 'crise de la quarantaine', a mid-life crisis, and in Catalan *fet bé* means *fait bien.* Altogether Séverin makes six wines from his own vineyards, as well as some bubbles from bought grapes. He is particularly enthusiastic about the white wines of the area, from limestone and from schist. He has vines in both terroirs: Grenache Blanc and Grenache Gris *complanté* on clay and limestone, and Macabeo on schist.

Engrenaches from 2017 is tight and firm, with incisive acidity and underlying oak, after an élevage in eight-year-old barrels acquired from Louis Latour. The wine is left on the lees, but there is no *bâtonnage.* In contrast, 2018 was more rounded. Makeba, as Séverin calls his Macabeo, is grown on schist, and also given some oak ageing. It is very saline and enjoyably austere, with good acidity. Macabeo is more acidic than Grenache, and is used for Vin Doux Naturel, when it would ripen at potential alcohol of 14–15% abv; Séverin prefers to harvest at potential alcohol of 12–12.5% abv. Yoshi is a dry Muscat à Petits Grains, from just 30 ares. Given some skin maceration, it has the discreet notes of the grape variety and some acidity on the finish. There is also a rosé, a blend of Grenache and Mourvèdre, but that sold out completely, so we could not taste any.

For red wine, Séverin favours what he calls a millefeuille of *égrappé* and *non-égrappé* grapes, with a proportion of the grapes pressed immediately. This is a technique that he has learnt from Jean-Philippe. There is no need for *pigeage*; it is an infusion, without any sulphur. Séverin talked about the virtues of racking, which gives good results if the wine is not filtered, and the racking is naturally protected by the carbon dioxide. Peaux Rouge is a blend of equal parts Syrah and Grenache, given a six-day *cuvaison*, after which it is aged partly in 600-litre barrels and partly in cement vats. The flavours were fresh and stony, with good, firm fruit. On checking the biodynamic calendar we discovered the day was a 'No' for tasting. I begged to disagree! Karignan spends seven months in demi-muids. The colour is deep and the nose quite sturdy; there is ripe fruit on the palate and a streak of tannin, with length, depth and an elegant finish, especially for a Carignan.

When I quizzed Séverin about the typicity of Calce, he enthused about the salinity and minerality of the white wines. They simply do not taste as though they come from a warm climate. The red wines too are fresh and elegant. Calce is quite different from the plain and there is an enormous variety of soils between the Pyrenees and the Corbières. If things work out, Séverin would like to extend his vineyard holding. He began working biodynamically in 2019, observing quietly, 'on va voir' – we shall see how it goes. If you are looking for somewhere to stay in Calce, Séverin has opened a gîte in the centre of the village. He deserves to do well.

Domaine Gérard Gauby

Calce

www.domainegauby.fr

It all began for Calce when Gérard Gauby took over his grandfather's vines, making his first wine in 1985. I first met Gérard in the mid-1990s when I was researching an earlier book and when I ignorantly asked him about the history of the estate, he proclaimed, borrowing a phrase, 'l'histoire, c'est moi'. His grandfather was a *paysan* who delivered his grapes, from just three hectares, to the local cooperative. His father was a government official, a *fonctionnaire*, who bought vines, so it is Gérard who has developed the family estate, initially with his two sisters. Gérard's success has drawn other aspiring wine growers to the area. Tom Lubbe arrived from South Africa and married Gérard's sister, Nathalie, and together they run Domaine Matassa, now based outside Cases-de-Pène. Thomas Teibert of Domaine de l'Horizon is a former son-in-law and others include Oliver Pithon, who originates from the Loire Valley, Jean-Philippe Padié and more recently Séverin Barioz.

I have visited Gérard regularly over the years. He is friendly and opinionated, and has not mellowed with age. On my last visit, the obvious question was: what has changed? He talked about the cooperatives losing steam, what he called 'une perte de vitesse', and of the explosion of independent wine growers, especially in Calce, Maury and Latour-de-France. It is they who have given energy to Roussillon, realizing its potential, he says. His own winemaking has changed too, though these days it tends to be his son, Lionel, who is responsible for work in the cellar, while Gérard concentrates on the vineyards, of which he now has

40 hectares, in numerous different plots. The style of the estate has undoubtedly become more elegant. The first wines of Gérard's that I would have tasted and drunk in the 1990s were mouth-filling and warming. Initially he fermented whole bunches, then he bought a destemmer, and now he and Lionel once again prefer whole bunch fermentation, which he feels gives more elegance and finesse.

You reach Gérard's cellars at the end of a very narrow track, off the road to Baixas just outside the village. A row of cypresses at the entrance reminded me of Tuscany. A tour of the vineyards is essential to observe Gérard's work, even in the unusually damp mist of a June morning. Gérard talked about organic agriculture, not just viticulture. For Gérard there has been a continual evolution. Initially he followed the conventional viticulture of his grandfather, and from that he progressed to *lutte raisonnée*, then organic and now biodynamic viticulture. In the cellar they use as little sulphur as possible, and natural yeast rather than the selected yeast that he favoured at the beginning of his career. He does not want to fit into any category: 'it is not important,' he says. 'What is important is to work *dans la purité*.' He favours stainless steel over cement vats, for the simple reason that you would use about 3 hectolitres of water to clean a 30-hectolitre cement vat, whereas one bucket of water is sufficient for a stainless steel vat.

Our discussion continued over a tasting of a pair of 2018 Calcinaires, Côtes Catalanes. The white, from Grenache Blanc and Macabeo, with a little Vermentino and Chardonnay, was very intriguing. It was quite golden in colour, with a rounded, herbal nose and quite a rich palate, with good acidity, a long finish with good tension and notes of salinity and intruding nuances. The wine underwent a malolactic fermentation and the blending took place as early as possible. Gérard no longer makes any appellation wines, after a 2014 was too low in alcohol for the appellation and he was told to rectify the 'problem' with concentrated must. He was, needless to say, outraged. Red Calcinaires is roughly half Mourvèdre, with some Syrah and tiny quantities of Grenache, Carignan, Cabernet Sauvignon and Cinsault, blended early and aged in foudres and vat. It had elegant, spicy fruit, with some leathery notes, some supple tannins and a long lingering finish.

Gérard's cuvées follow the pattern of Burgundy, with a village wine, a *premier cru* and then a *grand cru*. The youngest vines are used for Calcinaires, the village wine. The *premier cru* is the Vieilles Vignes, with Carignan and Grenache vines of at least 60 years old and some

35-year-old Syrah. The very oldest vines are used for Muntada, the *grand cru*, made from 90-year-old Grenache and 130-year-old Carignan. Alcohol levels are modest compared with much of Roussillon; 12.5–13% abv for the red and 12% abv for the white wines. Lionel has added other wines to the range, including Coume Gineste, both red and white, La Foun from old Carignan and La Roque from old Grenache.

Gérard's parting thought was: imagine the agriculture of the future without water. Irrigation is impossible; there simply is not enough water, with so many tourists coming to the region. However, if the department was to plant a million trees over the next ten years, they would act as windbreaks. There would be more rain, and the trees would absorb the pollution.

Domaine Padié

Calce

www.domainepadie.com

Jean-Philippe Padié has his cellar in the main street, in the heart of the village. His first vintage in Calce was 2003. Before that he worked at Mas Amiel for a couple of years and then with Gérard Gauby. He is enthusiastic about Calce and says that there is something magnetic about the terroir. In contrast to Maury, which is closed in, in a valley, here you have horizons, the hills of the Corbières and the snow of the Pyrenees, with a variety of soils. Gérard was the first to understand that. Jean-Philippe comes from a farming family near Chartres, but grew up near Burgundy. He admitted that initially he was not very interested in wine, but studied agriculture, which led to wine, and the rest is history. Jean-Philippe began with six hectares and now farms 18, half of which he owns, and the other half of which are rented, *en fermage*. His vines vary in age from brand new to 100 years old, with a lot of 60- to 70-year-old vines.

He makes an eclectic range of wines, all Vin de France. The whites were originally Côtes Catalanes, but he did not feel that any of the appellations reflected his wine. For him communicating his terroirs sincerely is important and outweighs the disadvantages of being outside the appellations, which he considers in any case to be more suited to larger producers and retailers. He has gained in liberty, and is excited by the changes he sees in Roussillon. 'The white wine has enormous potential,' he says, 'with wines like l'Argile from Domaine de la Rectorie and Gérard Gauby's Coume Gineste. Roussillon is waking up, thanks to the white wines, and *rancio* too is following.'

His cellar is full of foudres, barriques, and cement and stainless steel vats. As we tasted, we watched the world go by through the open cellar door. The village mayor and the ex-mayor, on their morning constitutional, popped by to say hello. My tasting began with Calice, its name a play on Calce, the entry level wine from younger vines. From Grenache, Carignan and Mourvèdre, grown on schist, it had fresh fruit, crunchy cherries, and a streak of tannin. Jean-Philippe called it the Beaujolais of Roussillon, observing that you can find similarities between Carignan and Grenache Noir. Both are *mal aimés*, or unloved, and both need a terroir that limits them in order to obtain interesting results.

Jean-Philippe favours a whole bunch fermentation and infusion of the grapes. There is no forced extraction, no *pigeage* and no *remontage*. A third of the grapes are pressed and then the whole bunches are added; the free-run juice and the pressed juice are blended at the end of the fermentation. It is all very gentle. After Calice, came Petit Taureau, taking its name from a song by Claude Nougaro. It is a blend of 50-year-old Carignan vines, with 30 per cent Syrah and a little Grenache and Mourvèdre, from both limestone and schist soils, what Jean-Philippe called yin and yang. He believes limestone is more masculine and schist more feminine, with solar and lunar energy. Here the infusion is longer but the method the same. Jean-Philippe explained how he used to crush the grapes, but in 2014 he did not have enough space, so he simply put the Carignan in the press and added the juice when there was room.

The 2018 Gibraltar followed, also named after a song, this time by the rapper Abd al Malik, and the successor to Ciel Liquide, named after a Serge Gainsbourg lyric. (My knowledge of French popular music was on a sharp learning curve that morning.) This is pure Grenache, from 50-year-old vines grown on blue and grey marl. It was fresh and fragrant, with Jean-Philippe observing that he saw it as the style of modern Grenache, more refreshing with some elegance. The 2015 followed the same pattern.

Next came Le Tourbillon de la Vie, named after a song from the iconic Truffaut film *Jules et Jim*. The wine is a selection of Macabeo, from granite in the Haut Fenouillèdes around the village of St Martin, and also from limestone at Tautavel and Opoul. Jean-Philippe is looking for refreshing white and the flavours were nicely floral, with iodine notes and a dry finish. Although he favours natural wine, he will add a little sulphur at bottling. 'I am not a fanatic!' he insisted.

While Jean-Philippe does not give his white wines any skin contact, they tend to be quite golden in colour. Fleur de Cailloux, which was a golden orange, comes from Grenache Blanc, with 40 per cent Macabeo and 10 per cent Grenache Gris, from both young and older vines grown on a mixture of soils. It was aged in both foudres and stainless steel vats. Grenache Blanc tends to have deep-coloured grapes, and the palate was dry and textured, with salty mineral notes and good acidity.

Milouise is named after his great-grandparents, Milo and Louise, whose photograph hangs in the cellar, with Jean-Philippe as a very young child sitting on Milo's lap. Milo was a farmer and had a vineyard of Chasselas de Moissac. The vines for this cuvée are *complanté*, from different terroirs all around the village, and what Jean-Philippe called 'tutti frutti'. Grenache Gris is the main variety, but there is also Macabeo, Grenache Blanc, Tourbat, Clairette, Carignan Blanc and both Muscat d'Alexandrie and Muscat à Petits Grains. The grapes are pressed and immediately put into foudres. The colour was golden, the palate firm and salty and the wine had good tension and a certain weight. The 2015 Milouise followed the same line, with good acidity.

Jean-Philippe has planted some Cinsault on schist – we enthused about the wines of Thierry Navarre in St Chinian – and has also planted Vermentino as he loves the white wines of Patrimonio in Corsica. This prompted him to open a bottle of Clos Marfisi which was a new producer for me, and quite delicious. Jean-Pierre expects his Vermentino, from a vineyard surrounded by old Macabeo, to come into production in 2025. He has also planted more Grenache Gris and Carignan Blanc, observing that they can be difficult to find, but: 'We must keep the old vines and return to the old grape varieties, and continue the Roussillon tradition of field blends.'

Domaine Olivier Pithon

Calce

www.olivierpithon.com

Olivier Pithon comes from the Loire Valley; Jo Pithon, of the eponymous wine estate in the Coteaux du Layon, is his brother. The obvious question to ask Olivier was why he came to Calce, and the reply was a mixture of opportunity and encounters. He never thought he would be able to be a vigneron. Vineyards are too expensive to buy. At least they certainly were in Bordeaux, but not, Olivier realized, in Roussillon. He came here looking for work, after studies in the Loire and Bordeaux, met Gérard Gauby in November 2000 and has never left. He loves the region: it was very welcoming, and he found 7 hectares of vines and a house and cellar in the centre of the village. His first vintage was 2001 and he has gradually increased his vineyard holding to 25 hectares, mostly on the road to Estagel. In 2015 he moved into a spacious new cellar on the outskirts of the village. For red wine, there are numerous concrete vats, which Olivier particularly likes as the temperature does not change and that maintains the fruit. For white and rosé, he prefers stainless steel vats. At the end of the cellar he has dug into the rocks, so part of it is underground with foudres, tronconic vats and demi-muids. All his wines have been biodynamic since 2010. In addition, he has developed a négociant business.

Olivier has eight different grape varieties in three or four different terroirs. Schist is the best for white wine, giving freshness and acidity. He loves the freedom of Roussillon: 'You can do anything; there are no barriers,' he says. Olivier's range is very logical and with one exception, all his wines are Côtes Catalanes. He does not think the appellation really means anything to anybody.

I spent a happy hour or so tasting with Olivier and his assistant, Anaïs. Mon P'tit Pithon Blanc comes from bought grapes that are organic, but not biodynamic. The white is a blend of one-third each of Macabeo, Grenache Gris and Grenache Blanc, with some sappy, herbal fruit. Cuvée Laïs, named after the estate cow, who is almost as old as the domaine, includes more Macabeo, with 20 per cent each of Grenache Gris and Grenache Blanc. Most of the blend spends ten months in oak and the oak is nicely integrated with fresh acidity and salinity. D18, named after the number of the road towards Baixas that crosses Calce, is a selection of Grenache Gris and Grenache Blanc from the best plots of schists that spent 14 months in 20-hectolitre foudres. The vines are 30 to 40 years old. The young wine was solid and intense with great ageing potential. Maccabeu is a varietal wine, from a grape variety that Olivier particularly likes. Grown on schist and aged in barriques, it has structure and firm acidity. Olivier observed that it is rare to find Macabeo as a pure varietal.

Mon P'tit Pithon Rouge, a négociant wine, is a blend of equal parts Grenache, Syrah and Mourvèdre, kept in concrete and stainless steel vats. The aim is easy fruit, and that is just what Olivier has achieved; cherry fruit with a refreshing finish. Cuvée Laïs Rouge is a Côtes du Roussillon, from equal parts Carignan and Grenache plus 20 per cent

Mourvèdre, aged in concrete vats and foudres to give some smoky fruit and structure. Pilou, the name of the plot, which is barely one hectare, is pure Carignan, from vines that were planted in 1898, just after the phylloxera crisis, with an additional planting in 1917. The flavours are ripe and fresh with stony acidity and structure. Le Colt comes from a plot of Grenache Noir in the middle of some white varieties. Olivier had to change the name from Clos, as the term *clos*, or *clot* in Catalan, can only be used for an appellation, which demands three grape varieties, and the vineyard must be walled, and this is not. It has been aged in foudres, and has some firm red fruit and structure.

Olivier observed that there is no history of ageing wine in Roussillon, unlike the Loire Valley which appreciates what he called the culture of the vintage. To prove that Roussillon can age, he opened a couple of older bottles. First came Le Pilou from 2009, a very hot, dry year, with drought conditions lowering the acidity. It had firm, leathery fruit on the nose, a cedary palate and certainly did not taste as though it came from a hot year. Then Olivier opened D18 from 2010, which was a cool, fresh year and a complete contrast. Apparently, the wine was not very expressive in its youth, but nine years on it had developed some intriguing dry honey and some herbal notes. Mature Chenin Blanc came to mind.

Olivier is not worried about his acidity levels. He does not want to pick too early and does not want his Grenache as low as 12–12.5% abv, instead he looks for a combination of freshness and generosity. He would like to plant some Cinsault, as he likes its freshness and elegance, but is not interested in Chenin or Riesling. He is also replanting some Grenache Gris and has planted some Carignan Blanc, another overlooked grape variety. For Olivier, 'Roussillon is traditionally a region of *assemblage*, of blending; the purity of a single variety can be fabulous, but it is possibly less complex. Everything is possible in Roussillon.' However, he is not interested in Vin Doux Naturel.

Les Vignerons du Château de Calce

Calce

www.chateaudecalce.com

Valérie Balmigère is a friendly woman who shares the presidency of Les Vignerons du Château de Calce, the name of the Calce cooperative, with Maximilian Paillard, whom I did not meet. She explained that the cooperative was founded in 1932 by her great-grandfather. At the time, most people in the village were shepherds, with some vines, and everyone made 'their own wine in their own corner'. Six founding cellars took the decision to join up and share equipment. Today there are 160 hectares in Calce and the surrounding villages, with about 30 *viticulteurs*. The cooperative bottled its first wine in 1980, but until fairly recently its real focus was wine *en vrac*. In the past they sold a lot of Muscat to the large French spirits group La Martiniquaise, and even today half the vineyards are planted with Muscat. The rest is Syrah, Grenache and Carignan, some Grenache Blanc and Macabeo and just one plot of Carignan Blanc.

Valérie's father was president of the cooperative when they won the case that allowed cooperatives to use the term 'château' on their label – this was a milestone piece of legislation. Hitherto it was only allowed for private estates. The case went as far as Brussels in 1994, following hearings in Perpignan, Montpellier and Paris. Other cooperatives have followed suit, notably La Chablisienne with its label, Château de Grenouilles. The château itself is at the top of the village. The first written reference to it is from 1232; it has had a chequered history and has belonged to various powerful families. Today it has been much restored and is now divided into private homes.

For Valérie, the typicity of Calce depends on the foothills of the Corbières, making for wine with good acidity, freshness and crunchy tannins. The terroir of schist and limestone provides minerality. Their range of wines, in contrast to the rest of Calce, is very classic. A white Côtes du Roussillon is a blend of Grenache Blanc, Macabeo and, since 2018, Vermentino, with some herbal fruit. Calcidoine Blanc is a blend of Grenache Blanc and Macabeo kept in barrel. There is a pure Carignan, Côtes Catalanes, including some 100-year-old vines, with some fresh concentration. They use both carbonic maceration and a classic vinification. It is named in memory of Séverin Estirach, described as a great defender of the wines of Roussillon; his granddaughter is a member of the cooperative. The Côtes du Roussillon Rouge is a blend of Syrah, Grenache Noir and Carignan, with some berry fruit, as is the Côtes du Roussillon Villages. Calcedoine Rouge in contrast, is aged in wood for 12 months and has vanilla and structure.

Next came a classic range of Vins Doux Naturels, first a Muscat de Rivesaltes, mainly from Muscat à Petits Grains, and then a Muscat Vieilli en Fût, with the oak changing the profile completely and adding another dimension. A 2009 Hors d'Age Ambré, aged in vat, was rounded and nutty. The 1996 Hors d'Age, aged in barrel for two years and then in vat for ten years, was quite austere with some underlying richness. Rivesaltes Rosé exists as a category of the appellation but is quite complicated to make. In addition, there is little communication about it, so it tends to be overlooked. Here, it was light and fresh with red fruit and a touch of citrus. We finished with a 2010 Rivesaltes Tuilé with spicy berry fruit.

I sense that the position of a cooperative in a village of successful wine growers, each with a high public profile, is far from easy. The average age of the members is rising, and many have no children to take over their vines. Its future may be uncertain but for the moment Valérie is an effective co-president.

Above: View of the village of Maury, with the Corbières hills behind.

Below: The impressive *parc des bonbonnes* at Mas Amiel.

Above: View of the snow-capped Canigou.

Below: Geological map of Roussillon.

- Red soils on compact limestone
- Limestone soils on the edge of the Corbières and Thuir
- Black soils, schistous marl from the Cretaceous Period
- Soils seated on schistous colluvial deposits
- Soils originating from granite and gneiss sand
- Very stony soils on high terraces
- Very stony soils on low terraces
- Clay and silt soils of Les Aspres
- Stony clay soils of Les Aspres
- Stony soils of the Thuir terraces
- Sandy, stony soils of the Albères foothills
- The whole range of stony, sand and silt, clay and silt, and sandy soils of the coast
- The hydromorphic soils of the basins
- Limit of the permanent salt marsh

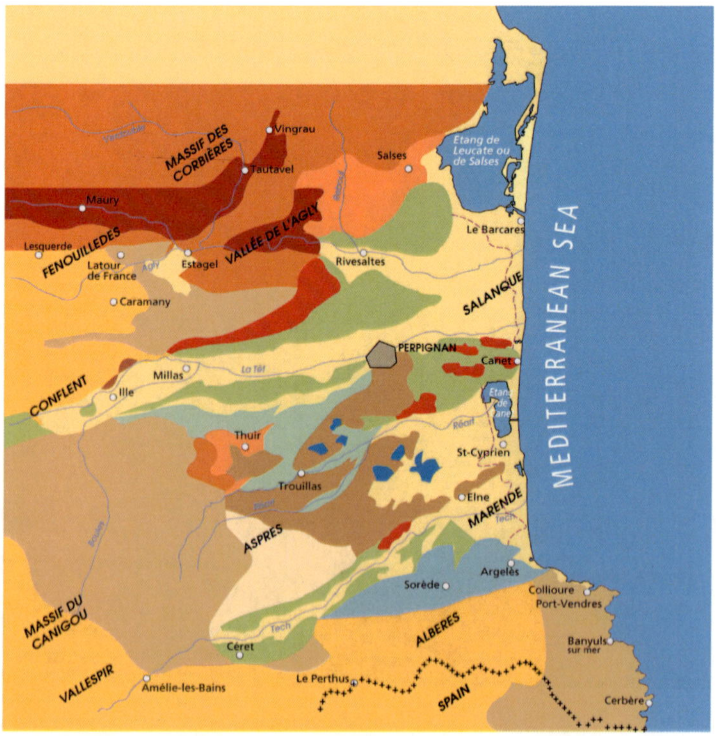

Right: Bush vines at Tautavel with the dramatic cliffs behind.

Below: The undulating vineyards of Les Aspres, near Terrats. Note the red, iron-rich soil.

Left: An old Grenache Noir bush vine, pruned *en gobelet*, with a generous yield.

Below: Bush vines coming to life in the spring. Note the stony terrain.

Above: Vines to be planted. The graft is protected by wax.

Below: Gulleys caused by erosion.

Above: Vines planted *en echalas*, with a wooden stake for each vine. On the hill behind is the watchtower of Madeloc.

Below: The dramatic Côte Vermeille, with vines almost falling into the sea.

Above: Terraced vineyard in the foreground, with Banyuls-sur-Mer in the background.

Below: Steep, terraced vineyards in Banyuls, with stocky bush vines on very stony soil.

Above: Coastal vineyards near Banyuls.

Below: Looking down on the harbour of Collioure from the vineyards above.

8

CASES-DE-PÈNE TO TAUTAVEL

CASES-DE-PÈNE

Château de Jau

Cases-de-Pène

www.chateaudejau.com

I first visited Château de Jau back in the last century when Estelle Dauré ran the estate. These days Estelle has gone on to pastures new and it is her brother, Simon, who is very much in charge. This is an old estate, going back ten centuries, when it was the Cistercian abbey of Jau. The remains of a twelfth-century tower still stand. There are periods of its history about which little is known, what Simon called 'zones ombres', or shadowy periods, but he supposed that the monks had disappeared during the French Revolution. Various owners had run the estate before the Dauré family bought it in the 1970s. At that time his father was a négociant, specializing in Vin Doux Naturel, who dreamed of making his own wine. Simon related how the *crise de la quarantaine*, a mid-life crisis, made his father realize that he wanted to return to the land, so he sold the family négociant business in Perpignan and came to run Château de Jau.

Simon took me for a drive through the vineyards. The estate totals 100 hectares of vines as well as extensive *garrigue*, with a variety of different soils – sand, schist, limestone – with most of the vineyards in the valley. There has been some extensive replanting as they work towards organic viticulture, and HVE, with bee hives, and sheep grazing in the winter. Simon considers himself lucky to have access to water, and flat vineyards, so irrigation is feasible. His average yield is a generous 45 hectolitres per hectare. He also has olive and almond trees, and even produces a Muscat de Rivesaltes vinegar. Red wine dominates the production, with 30 per cent white and just 5 per cent Vin Doux Naturel. When his father bought the estate in the 1970s, it only produced Vin Doux Naturel. In contrast, these days they have just 50 ares of Muscat remaining. 'We have turned an enormous corner,' observed Simon. 'Vin Doux Naturel is a fantastic heritage, but with a niche market, making it so difficult to sell.' He considers that 'the climate is our trump card; it really helps for organic viticulture. We have wind and sun and are in a valley, with no neighbours.' Most of the vineyards are mechanically harvested, with Simon ruefully commenting that it is not politically correct to enthuse about a mechanical harvest, but they are now so efficient.

They make an extensive range of wine, of which Simon gave me a selection to taste. La Jaja de Jau, a brand that was created 35 years ago, has a distinctive hand-written label by a Swiss artist. A Pays d'Oc Chardonnay, from bought grapes, was fresh and rounded; Simon prefers Pays d'Oc to Côtes Catalanes, observing that Côtes Catalanes is less visible. Château de Jau is a blend of equal parts Vermentino and Macabeo, with a touch of oak; it is fresh and pithy. Simon would like other varieties included in the appellation, such as Chardonnay, but Sauvignon Blanc is not worth the bother. He does not think that the appellation helps sales, but if your wine is not an appellation, you cannot use the term 'château'.

Jaujau 1er blanc, with a first vintage in 2017, is a blend of Vermentino with Grenache Blanc and Grenache Gris, aged in new 500-litre barrels for 12 months – the oak was well-integrated and the wine had good acidity. Le Jaja de Jau Rosé, Pays d'Oc, is a blend of Syrah and Grenache. 'Syrah is complicated; the skin contact must be the absolute minimum,' says Simon. Le Jaja de Jau Rouge is a pure Syrah, with easy drinking fruit.

A no-added-sulphur Côtes du Roussillon Villages called 0%, with a dramatic label, has rounded fruit. It is based on Syrah, with some Grenache and Carignan. Simon is replanting Grenache and Syrah, which works well in his vineyards, but not Carignan, which he finds too sensitive. His vineyards are not too hot and have good ventilation. He irrigates if necessary: you have to ask for a derogation, but it is usually allowed.

Château de Jau Côtes du Roussillon Villages is the heart of the range, based on Syrah with 20 per cent Grenache and 10 per cent each of

Carignan and Mourvèdre, kept in vat for 15 months and blended after ageing. Syrah is picked first, then Grenache, followed by Carignan and lastly Mourvèdre. The flavour was ripe and spicy, supple with a streak of tannin, making a very appealing glass of wine.

Jaujau 1er Côtes du Roussillon Villages is a blend of Syrah and Grenache, with more concentration and, to my taste buds, less charm. The tasting finished with a sparkling wine, a blend of 70 per cent Colombard with 30 per cent Muscat, which was grapey and frothy, reminding me of Moscato d'Asti. Simon had first made a pure Muscat, but that was 'un peu too much'. This was lightly sweet with 80 grams per litre of residual sugar and 5% abv, bottled with a screw cap, providing a fun picnic wine. You sense that Simon really enjoys marketing.

He also has a property in Chile, having bought land at Apalta in the Colchagua Valley over 20 years ago, making Viña las Niñas. He described it as a great adventure; commenting that Chile is a land of milk and honey for wine.

Château de Péna

Cases-de-Pène

www.chateau-de-pena.com

Jean-Christophe Bourquin is president of Château de Péna, the small cooperative of Cases-de-Pène. He and his vivacious assistant, Céline Preschey, took me for a drive through the vineyards, which are on gentle slopes, the first foothills of the Corbières, behind the village as well as on the plain towards Espira-de-l'Agly. They pointed out the Ermitage Notre-Dame de Pène, a small chapel on a hill above the village. There are 54 steps up to it, which I have yet to climb. This small cooperative, founded in 1942, has 35 members cultivating a declining 320 hectares. Just ten make a living from their vines. All practice sustainable viticulture and are working towards HVE and even organic viticulture. Biodiversity is important, with alternate rows planted with grass; the soil is analysed to see what elements are missing and they are concentrating on site selection. The cooperative's cellars, on the edge of the village, are well-equipped and there is a welcoming tasting area in the shop.

These days the ratio of Vin Doux Naturel to vin sec produced here is 15 per cent to 85 per cent, compared to half and half when Jean-Christophe became president in 2000. Muscat de Rivesaltes accounts for most of the Vin Doux Naturel and they make very little Rivesaltes. They describe the house style as wines for early drinking, ripe, with silky tannins. Les Affranchis de Péna (*les affranchis* means rebels, or free spirits) is the entry level range: the white is a blend of Macabeo and Grenache Blanc, with a hint of oak and some white blossom. Côtes du Roussillon Les Pierres Noires Blanc is almost pure Grenache Blanc, with a hint of Macabeo, fermented and aged in oak. Ninet de Péna Blanc, Côtes Catalanes, is pure Muscat à Petits Grains, with fresh, grapey fruit (*ninet* is an affectionate term for a small boy). Ninet Rosé is a cheerful *rosé de piscine*, while Ninet Rouge is an easy blend of Grenache and Carignan, with some fresh spice. The red Les Affranchis is a blend of equal parts Grenache, Mourvèdre and Syrah with some leathery spice. Côtes du Roussillon Villages Château de Péna, a blend of 40 per cent Syrah, 30 per cent Grenache, 20 per cent Mourvèdre and 10 per cent Carignan is unoaked, with rounded spice and supple tannins. Les Pierres Noires, in contrast, has spent 12 months in wood, and comes mainly from Grenache with Syrah and Carignan.

Then we went on to the Vins Doux Naturels. Céline explained that the Cuvée Regards de Femmes, Muscat de Rivesaltes, is blended by a large group of women who do not necessarily work in wine, and raises money for a local cancer charity Equilibre 66. Pas d'Excuses, Côtes Catalanes, is a less unctuous Muscat, intended for those who say they 'do not like Muscat as it is too sweet'. As the name implies, this wine allows for no excuses. It is fresh, with some dry honey, and unfortified, so the alcohol level is a refreshing 12% abv. The classic Muscat de Rivesaltes is ripe and more honeyed. A Rivesaltes Tuilé Hors d'Age, a pure Grenache Noir, kept for about 15 years in wood, was elegant, with dry, nutty notes, balanced with good acidity. Rivesaltes Ambré, a blend of Grenache and Macabeo, also aged for about 15 years, had dry apricot fruit. Céline lamented the 'has-been' aspect of these wines, that they are no longer in fashion.

ESPIRA-DE-L'AGLY

Mas Crémat

Espira-de-l'Agly

www.mascremat.com

I first visited Mas Crémat for an earlier book, when Jean-Marie Jeannin talked about their arrival in Roussillon. He and his wife came from Burgundy; she is part of a well-known wine family, Mongeard-Mugneret, and he had worked in the dairy industry. They acquired Mas

Crémat in 1990, with 30 hectares of vines, for a price equivalent to one hectare of Vosne-Romanée *premier cru*, assuming that you could have found any for sale. The farmhouse was in ruins and the grapes went to the cooperative. It is an old property, built in 1800, and a concrete vat in the cellar is dated 1865. Today it is their children running the estate, Julien Jeannin and his sister Christine. Julien's first proper vintage was 2006, after studies in Bordeaux.

The vines are mostly close to the cellars outside the village of Espira-de-l'Agly. Julien took me for a drive through the vineyards, which are on terraces going down to the river, with a view of the Romanesque church tower of Espira-de-l'Agly in one direction and the watchtower of Tautavel in the other. The IGP vines are by the river; they resorted to irrigation in 2019, so that the grapes would ripen properly. During the heatwave, the temperature reached 55°C for four hours, perhaps appropriate considering the property name – *crémat* means burnt in Catalan. They are working on *enherbement*, planting cover crops such mustard and various varieties of beans every other row, which will help the structure of the soil. The crops are ploughed in every spring.

They make an extensive range of wines, some 35 different cuvées in all, rather like Burgundy, observed Julien. I tasted a small selection in their spacious shop. First come the IGPs, red, white and rosé, and a Muscat Sec, next are the classic appellations and then some Vin Doux Naturel, 'pour s'amuser' – for fun. His parents, on the other hand, never made Vin Doux Naturel. There are two special cuvées, with their names relating to young family members: Les Sales Gosses, little boys who are cousins born two months apart, and Les Petites Demoiselles, again cousins, who are two weeks apart. With Les Sales Gosses, they do something different each year. I tasted the 2018 which was made without sulphur.

Côtes du Roussillon Blanc, La Yose, meaning schist in Catalan, comes from 60 per cent Grenache Blanc with some Vermentino, which gives acidity and balances the rounded texture of the Grenache. Although the whole estate is classified as Côtes du Roussillon Villages, Julien considers that the IGP allows for more diversity in his range. Côtes du Roussillon L'Envie is a blend of Grenache, Syrah, Mourvèdre and Carignan, with the precise blend varying with the vintage. Grenache gives rounded fruit, Carignan acidity, Syrah spice and Mourvèdre tannins. Julien wants what he called the 'côté digeste', and the wine was rounded and fruity, with a supple finish. Cuvée Bastien, a Côtes du Roussillon Villages named after his mother's first grandson, and not one of the *sales gosses*, is a blend of equal parts Syrah, Grenache and Mourvèdre, aged in barrel, with some weight and density.

Rivesaltes Grenat, pure Grenache Noir, was muted *sur grains*, with ripe, spicy fruit and light tannins. His Muscat de Rivesaltes was muted at 100 grams per litre of residual sugar; he says that below 100 grams per litre you do not get the right balance of sugar and acidity. It was deliciously perfumed. Asked about projects for the future, Julien said 'holidays', but behind his flippant response and relaxed approach is a firm commitment. He is replanting vineyards, would like to work on biodynamics, and is considering eggs for élevage. He may also plant some almond trees. The change of generations has been successful.

Domaine Danjou-Banessy

Espira-de-l'Agly

04 68 64 18 04

Sébastien Danjou is a welcoming host. He talked about his family history and how it all began with his great-great-grandfather, who had vines and sheep, as well as almond and olive trees. Polyculture was the order of the day. It was his great-grandfather who concentrated on the vineyards and bought the property where we met, on the outskirts of Espira-de-l'Agly, which had once been the old stables of the abbey of Espira-de-l'Agly. At that time you could make a living with five hectares of vines. His parents had done other things, and encouraged Sébastien and his brother Benoit to study; Benoit opted for French literature, and Sébastien English literature, working for a time as an English teacher. But when in 2001 their grandfather announced that he was going to sell the vineyards, they said, 'No, you can't'. They may not have studied viticulture or winemaking, but it was in their genes, and both had worked in their grandfather's vineyards as children. He had concentrated on Vin Doux Naturel, while his grandsons today make vin sec, *au feeling*, as Sébastien put it, without any formal oenology studies. They are also very enthusiastic about *rancio*, with Sébastien observing that it goes back to antiquity whereas Vin Doux Naturel was developed during the Middle Ages.

The brothers now have 17 hectares of vines, all around Espira-de-l'Agly and also towards Vingrau, on limestone, clay and different schists. The only thing missing is granite. In some places you can find limestone and schist mixed up together. They have replaced old vines that were no longer productive, and grafted vineyards to change their composition. They favour massal selection for replanting – their vineyards have an average age of 60 years – and are certified organic. They use some biodynamic preparations, with Sébastien observing that biodynamics work well in a damp climate, but not in the dry Mediterranean climate, where the soils really 'live' between October and March. He considers the moon to be an important influence in the growing cycle. However, he emphasized the importance of the soil, as that is what makes wines taste different, and they mention the soil type on their labels. All their wines are Côtes Catalanes.

Although the first vintage of the new regime was 2003, Sébastien explained that they started afresh in 2010, with a *tabula rasa*, or clean slate, to look for freshness in the vineyards. Phenolic maturity in Roussillon is at around 16% abv potential alcohol, which is good for Vin Doux Naturel but not vin sec. They wanted less concentration, still with ripe grapes, but not at 16% abv. They favour minimum intervention, crushing the grapes for an infusion, but no *pigeage*. 'The nose and the glass are the most important winemaking tools,' says Sébastien.

We settled down to a comprehensive tasting. Coste, Terrasses de l'Agly, grown on clay and gravel, with a limestone subsoil, is a pure Macabeo, from 40-year-old vines. The wine is kept in 20-hectolitre foudres, usually for 13 to 14 months but there is no fixed rule. The malolactic fermentation also depends on the year although in general they would prefer it not to happen. The wine was taut with incisive acidity and stony fruit. Sébastien will use a little sulphur if necessary, observing that if he has a headache he takes an aspirin.

La Truffière Blanc comes from Carignan Gris grown on black schist, with some limestone. It spends 18 months in barrel, making for some firm fruit and a satisfying mouthfeel. Originally all their wines fitted into the appellation but they had administrative problems with Carignan Gris. Sébastien is fairly scathing about the bureaucrats who run the appellation system and all his labels say, appropriately, Vins Fins de Catalogne Nord. SuperNova Vin de France, an orange wine made solely from Muscat d'Alexandrie, with the Muscat flavours tempered by the skin maceration, has some refreshing tannins and a pithy orange flavour.

Roboul is a blend of younger vine Grenache Noir and Mourvèdre, given a three-week maceration and aged in barrels of varying sizes – 300-, 500- and 600-litre – for 12 to 14 months, with fresh fruit and supple tannins. Red La Truffière is a blend of Grenache Noir, Carignan and a little Mourvèdre, also with a three-week maceration and aged mainly in foudres, with the Mourvèdre in barrel. It was very elegant, with freshness, some cherry fruit and a balancing streak of tannin.

Les Myrs, a pure Carignan grown on black schist, had red fruit and a certain tannic structure with a fresh finish. Estaca, a pure Grenache from black schist, including vines that are 120 years old, has the appealing cherry fruit of Grenache with a streak of tannin. Sébastien admitted to an experimental amphora, as well as a barrique. Espurna comes from 100-year-old Cinsault vines grown on quartz. Aged in old barriques, it has silky fruit and elegant depth. Mirande is Syrah grown on a north-facing basalt slope. Sébastien explained that his grandfather planted Syrah in order to conform to the new appellation, when it was described as a *cépage améliorateur*, but what was it actually improving? The wine had some peppery black fruit after a long maceration and 12 months of ageing, first in barrel and then in concrete vat.

Before we left, Sébastien took us into the cellar to taste a couple of Rancio Secs. The 1952 was made by his grandfather. Although some is bottled, they still have some ageing in barrel. It had a green rim and was very powerful and intense, with incisive flavours. In comparison the 1949 seemed sweeter and very rich but equally intense. It is a vital part of Roussillon's history and happily now enjoying something of a revival.

Domaine Gardiés

Espira-de-l'Agly

www.domaine-gardies.fr

I enjoyed my visit to Domaine Gardiés, where I met young Victor Gardiés. An engaging example of the new generation of young wine growers, Victor is discursive, thoughtful, open-minded and passionate about his work. He studied viticulture and oenology at Montpellier and has worked in Bordeaux and Châteauneuf-du-Pape. At Paso Robles in California he learnt 'what he didn't want to do', although it was none the less a very interesting experience. Victor's father built their cellar near Espira-de-l'Agly in 2004 and they are now adding a 10-metre-deep storage facility. Altogether they have 35 hectares of vines in about 50 plots, half around their cellar and close to Espira-de-l'Agly, and half at Vingrau and Tautavel. The vineyards around the cellar are based on black schist, and are very suitable for Mourvèdre and Carignan, whereas those of Vingrau are higher in altitude, and based on limestone and clay. Victor took me to see them after our tasting. It was a magical spot in the autumn sunshine. Two distinct plots, Geneglas and Farines, are in a series of terraces and surrounded by walls. Ploughing is difficult and he uses a horse. The lower pH soils here retain freshness, making these plots very good for white wine.

In 1980 Victor's father was one of the first to make vin sec (his grandfather only ever made Vin Doux Naturel); 1994 was the first vin sec vintage that they actually sold. It is all relatively recent. Talking about his winemaking, Victor said, 'We need to search for what suits Roussillon. We can get freshness, but we must take care with extraction. The less we do, the more we keep the purity.' He favours infusions rather than extractions. His vineyards have been organic for 15 years and are in the process of conversion to biodynamics, but are not certified, with the observation, like Sébastien Danjou, that the biodynamic specifications need to take into consideration the differences in growing conditions between the north and south of the country. He feels strongly that organic viticulture should be the norm here since the wind keeps everything ventilated and helps prevent disease.

In the cellar the grey concrete vats for the red wines make an attractive contrast with the dull red walls. The underground barrel cellar houses barrels and small foudres. Victor is experimenting with barrels of various sizes, between 6 and 20 hectolitres to see what works best: 'Wood is *un monde à part*, a different world, and mastery of élevage in wood makes the difference between a good and a great vigneron. The difference between six and five hectolitres is enormous; five hectolitres is a barrique; six hectolitres is a foudre. French coopers usually stop at five hectolitres. It is a different way of working: a foudre is more reductive; it lengthens the flavour of wine, and polishes it.' Victor does not like wood for wood's sake, stating: 'The wine needs élevage, but the wood must not be noticeable, otherwise it is make-up.' He uses various coopers – Taransaud, Vicard, Seguin Moreau – and favours a light toasting or steaming. He observed that Italy has a longer tradition for foudres than France. Foudres or tonneaux will keep for 15 years, whereas a barrique needs replacing after three or five years.

Victor admitted that one of his aims is to make the best possible white wines. He encourages a malolactic fermentation, but rarely stirs the lees as that makes for too much body. He wants white wines with tension that will keep, and pertinently commented that you must not forget that wine is meant to be drunk. Les Glaciaires, a blend of Grenache Gris and Blanc, Roussanne, Macabeo and a little Tourbat, aged in foudres for eight to ten months, was quite golden with a discreet nose and a tight, stony, saline palate. Je Cherche le Ciel Blanc is a pure Grenache Gris, grown on black schist, which makes for more powerful wines than limestone. The wine is kept in foudres and had a rounded nose and palate, with a hint of orange petal and good weight. Les Clos des Vignes, from old vines of Grenache Blanc and a little Grenache Gris, grown in a vineyard surrounded by little walls, was rounded on the nose, with firm, stony, mineral fruit on the palate. Victor said that he liked tension and austerity and that is what you find in his wines. A pure Macabeo, Les Vignes de Mon Père, comes from a limestone vineyard at Vingrau in the shade of the cliff, so it ripens late. Fermented in 600-litre barrels and given 12 months of élevage, it was firm and tight-knit on the nose and palate, with an incisive finish. Victor observed that it is a mistake to pick too early, as it means you obtain more malic acid, which is hard without necessarily imparting freshness, and too much malic acid then produces too much lactic acid, which makes for fat, buttery flavours. 'And if you lees stir too much,' he says, 'you lose the *côté salivant*,' the mouthwatering freshness.

This is one of the few estates to have Tourbat and to make it as an almost pure varietal, with just 15 per cent Macabeo, in La Torreta. Everyone has pulled up Tourbat in favour of Macabeo, as it is fragile and suffers from *court noué*. But it is a late ripener, retaining freshness, so it gives tension while Macabeo fills out the middle palate. The wine had firm acidity, and stony fruit, with a long, restrained finish.

Les Millères was the first red wine, with Syrah and Grenache from Vingrau and Carignan and Mourvèdre from Espira-de-l'Agly. With freshness, fruit and structure, Victor deems it to be representative of the region. An entry level wine should represent the signature of the wine grower, 'un peu sa signature'. It had some appealing red fruit, with elegant tannins. Elevage takes place in a concrete vat and there is an infusion of the grapes. Victor does not crush them, and there is a percentage of whole bunches, with a maceration that last between eight and 18 days. In Je Cherche le Ciel Rouge, from Mourvèdre blended with a little Carignan, he is looking for freshness and spice, and uses a lot of whole bunches. It was elegantly mouth filling. He also makes a pure Mourvèdre, from low yielding 50-year-old vines. It was initially discreet, but evolved in the glass, with Victor observing that he does not know how to make what he called *vins immédiats*, that give instant pleasure.

Clos des Vignes comes from old Grenache and Carignan and has rounded cherry fruit. Les Falaises, which Victor described as representative of the domaine's typicity, comes from old Carignan grown at altitude, with some Syrah. They have used Syrah to replace the dead Carignan vines, and the two varieties are vinified together and given 18 months of élevage. The wine was rounded, elegant and complete with a long finish. La Torre, a pure Mourvèdre given 18 months in barrel, comes from nearby vineyards and has smoky red fruit and firm tannins. Victor considered that had it been blended, it would have lost its diversity and originality. Essentially, for his red wines he wants balance and for his whites he follows what he called 'the route of purity'.

When we moved on to the Vins Doux Naturels Victor enthused, 'Rivesaltes is a treasure.' The Vin Doux Naturel range consists of Muscat de Rivesaltes Flor, a Grenat Cerise and various vintages of Vieux Vin Doux Ambré. In a Muscat de Rivesaltes Victor wants a wine that is as fresh as possible. He mutes it at 105 grams of sugar per litre; if there is less sugar the alcohol can give an impression of sweetness, but the balance is complicated. If the sugar level is too low, the wine can be too heavy. His was fresh and lemony, with an elegant finish. A 1998 Vieux Vin Doux Rivesaltes Ambré, from Grenache Blanc and Gris with a little Macabeo, had been bottled 18 months earlier and was amber in colour, with rich, nutty, intense fruit, balanced with good acidity. Victor observed that the angels' share is very important; about 25 per cent is lost from the barrel and that concentrates the wine, enhancing the flavour. According to Victor, making Vin Doux Naturel is for *fainéants*, the lazy winemakers, for you put it in a barrel and simply forget about it. However, you do need a mother, the lees from a previous wine, and should never clean the barrel.

Domaine Matassa

Espira-de-l'Agly

04 68 64 10 13

Tom Lubbe is one of the pioneers of modern Roussillon, making some experimental wines in 2002, with 2003 his first official vintage. The name of his estate comes from the *lieu-dit* of a vineyard of Carignan in St Michel-de-Fenouillet and means tangle wood in Catalan and a skein (of wool) in Italian. Born in New Zealand, Tom learnt his winemaking in South Africa and speaks of his mentor, Louise Hofmeyr of Welgemeend Estate, with great respect, saying that he learnt restraint from her. Her wine was imported to England by Richards Walford, who also represented Gérard Gauby, and that led to the suggestion that

Tom visit Roussillon. The first person he met when he arrived in Calce was Nathalie Gauby, Gérard's sister and, as it turned out, Tom's wife to be. Tom gradually developed his own vineyards, first with a cellar in Calce, from where he moved to Montner before coming to Mas Ferriol in February 2019. He now has 20 hectares of vines, 10 in Calce and 10 in Espira-de l'Agly.

We talked in the courtyard under the shade of a large acacia tree. Tom has a lively mind, changing the subject, darting from one thing to another and although very engaging and articulate, is sometimes difficult to pin down. We covered a number of topics, including viticulture, whether to plough or not, cover crops, winemaking, orange wine, organic wine, and how one's palate can change. Tom commented that he now gets bored drinking white wine; I suspect that he has a low boredom threshold.

Altogether he makes around a dozen wines, and he opened a few as we chatted. Cuvée Marguerite is a blend of one-third each of Macabeo, Muscat à Petits Grains and Muscat d'Alexandrie, which spends anything between 14 and 30 days on the skins, making for a golden-orange colour. The nose was redolent of Muscat, with some acidity and tannin on the palate. Tom said that he was very committed to Muscat and has six hectares of it, 'but it can be bloody-minded'. All Tom's white grapes are fermented on the skins, which gives grip and tannin to a wine, what he called 'une jolie amertume'. He favours 'minimum intervention with maximum care'. The last vintage for which Tom used any sulphur before bottling was 2008.

As well as Muscat and Macabeo he has old Grenache Gris in Calce, some Cinsault on schist, Grenache Noir and Carignan and two small plots of Syrah. He has tried co-fermenting Syrah with Muscat. Next came Tattouine Rouge, from Grenache Gris, given one month's maceration. The colour was light red and on the palate fresh and juicy with a tannin streak. Tom made a comparison with Poulsard from the Jura, which has a similar light red colour. Then he commented that 'some people think that I am a little perverse in my winemaking,' observing quite pertinently that there is room for more delicate wines in the south. 'I refuse to accept that the sun is pushing us towards bigger wines,' he says.

Matassa Rouge is a field blend of Carignan, Grenache Gris and Grenache Blanc planted in the 1900s. Whole bunches are given a four-day maceration and the wine is aged in 500-litre barrels, with some elegant fresh fruit. Tom also uses fibreglass, amphorae and concrete vats, but no stainless steel as that leads to reduction. Our tasting finished with 2014 Cuvée Romanissa, the name of which means 'the place where the rosemary grows'. It is pure Lledoner Pelut, which Tom described as the Catalan Grenache, and was beginning to *pinotte*, with some intriguing fruit and leathery notes. It was labelled Vin de France as the alcohol was too low for an appellation, and people said it would not age. How satisfactory to prove them wrong.

The range also includes Olla Blanc, based on Muscat and Macabeo, and Olla Rouge, based on Grenache Noir and Macabeo. Cuvée Alexandria is pure Muscat d'Alexandrie, and Matassa Blanc is a blend of Grenache Gris and Macabeo. Tom has developed a new wine from Carignan and Mourvèdre grown in his vineyards in Espira-de-l'Agly, called Ace of Spades. You sense that he does not let things stand still. His next project is a Rancio Sec.

Domaine Piquemal

Espira-de-l'Agly

www.domaine-piquemal.com

I met Pierre Piquemal for a previous book but he is now retired and it is his daughter, Marie-Pierre, who runs the family estate. When I visited at the end of the last century they had a cramped cellar in a back street in Espira-de-l'Agly but they now have new cellars outside the village, with the rare luxury of space. Used for the first time in 2013, the new cellars have allowed them to fine-tune their methods and now that they are equipped with numerous small stainless steel vats they can work on site selection. As Marie-Pierre put it, 'Now we can refine our wines.' Where previously they had two large vats of Syrah, now they have five smaller vats.

They have 48 hectares in all, vineyards on schist at Espira-de-l'Agly and Cases-de-Pène, and on clay and limestone at Vespeilles near Salses, with *galets rouges* at St Estève. Marie-Pierre initially studied wine commerce, and only turned to winemaking after her brother's premature death in 2010. Her husband, Cédric, a retired parachutist, is retraining to obtain a qualification in viticulture.

They have an extensive range of wines from a variety of terroirs and vinify by plot and by grape variety. Pierre Piquemal planted Cabernet Sauvignon, Merlot and Viognier some 35 years ago, and those, along with Muscat Sec, are destined to be easy-to-drink IGPs. Les Terres Grillées Blanc, a blend of Grenache Blanc, Macabeo and Vermentino, is rounded and herbal. Originally it was oaked but, influenced by her white wine studies in Burgundy, Marie-Pierre decided to lighten it and to pick earlier, which made all the difference. The estate's very first bottling of Tradition was 1983, with the 2018 a blend of Grenache, Syrah and Carignan, making fresh, fruity accessible wine. Le Chant des Frères is a blend of equal parts Grenache and Syrah, aged in 400-litre barrels. Marie-Pierre commented that they use less oak than in their old cellar and she has just 30 barrels.

Les Terres Grillées Rouge, a Côtes du Roussillon Villages, based on barrel-aged Syrah with some Grenache and Carignan, has fresh youthful fruit and an appealing balance. Colline Oubliée, named for a plot that had been forgotten, and which they replanted with some Mourvèdre, Syrah, Grenache and Carignan, is given six months in barrel and has some firm, sturdy fruit. Galatée, based on Grenache grown on schist, with some Syrah and Carignan, was perfumed and elegant, while Pygmalion, based on Syrah with some Grenache and Carignan and aged in barrel, was sturdier with peppery fruit.

Marie-Pierre described Vin Doux Naturel as a world apart. It is complicated. She has an original take on Muscat de Rivesaltes, ageing it in barrel and selling older vintages, so that in the autumn of 2019 I tasted the 2003. It came mainly from Muscat d'Alexandrie, with some bitter-orange fruit on the nose, while on the palate it reminded me of Sauternes, with smooth, unctuous fruit. We tend to think of Muscat for early drinking, but it ages remarkably well. Like so many of the previous generation, Pierre Piquemal used to make a lot of Vin Doux Naturel to sell *en vrac*, but these days it is such a struggle to sell it in bottle.

VINGRAU

Domaine des Chênes

Vingrau

www.domainedeschenes.fr

When Alain Razungles' great-grandfather bought Domaine des Chênes in 1919, it was already an award-winning wine estate (having won a medal in 1912). Alain's grandfather then took over and was followed by Alain's uncle and father, but there was no room for Alain on the estate so he studied agriculture and oenology and became a professor at Montpellier. He took over the 35-hectare estate when his father retired, and for a number of years had two careers, as a vigneron and as a highly respected professor, working on aromas, with more than 160 publications to his name. He has particularly concentrated on the effect of climate change on aroma; it is not all negative, for example Cabernet Sauvignon now has fewer green pyrazines. One of his three daughters is interested in wine. Alain now farms 37 hectares, all around Vingrau and Tautavel, which he has converted to organic viticulture. He has never used insecticides or pesticides; it is a question of controlling the grass.

Vingrau is part of the *cru* of Tautavel. They called it Tautavel as the name was already well-known for the prehistoric man, found by the *curé* of Vingrau at the end of the 1960s. Alain enthused about the terroir, saying that Roussillon is the most complicated of terroirs, for it has everything except volcanic rock. It can produce white wine with a wonderful freshness, and reds from clay and limestone, with the *côté tannique*, making them *vins de garde*. Vingrau has a lot of *galets roulés*, and the soil is dominated by clay and limestone. It is very stony, thanks to *éboulis calcaires* from the nearby cliffs, with numerous springs fed by the water table, so that Vingrau fears drought less than some areas. '*Tout est bon, comme dans le cochon*,' said Alain, which translates as everything is good, like in the pig! Altitudes vary between 110 and 400 metres. There is clay and limestone in the valley, which is surrounded by cliffs. Part of the vineyard is on *grès*, or sandstone, which is the parent of schist. However, wines from those vineyards are quite different. The cliffs provide shade so some vineyards do not get the sun until 10 o'clock in the morning. For Alain, Vingrau is particularly good for white wine, benefiting from this shade, with the limestone giving minerality and freshness. Some of the vineyards for the white varieties are north-facing, so only enjoy short periods of sunshine. Tautavel also enjoys the influence of the river Verdouble, a tributary of the Agly. With some schist in vineyards closer to Maury, it all makes for wonderful blending possibilities.

We tasted in his small cellar on the outskirts of the village, which does not even boast a café to its name. His white Les Sorbiers Vieilles Vignes comes from Grenache Gris and Grenache Blanc, with some Macabeo, all *complanté* with some Grenache Noir – it is like a garden, observed Alain. It is aged in old wood, and has a firm, mineral palate and good acidity. Les Magdaléniens, Côtes du Roussillon, a blend of Roussanne and Grenache, grown in the middle of the Cirque de Vingrau, is fermented and aged in wood and has a firm resinous nose and a more honeyed palate. Alain observed that he discovered Roussanne with Gérard Jaboulet's father when he was studying Syrah in the northern Rhône.

The name Magdaléniens comes from an encampment of early inhabitants of the area; they were hunters and fishermen 20,000 years ago and left behind harpoons and arrow heads.

L'Air du Temps Côtes du Roussillon has powerful, ripe fruit that almost jumps out of the glass. Les Grands-Mères is mainly old Carignan; the youngest vines are 55 years old. It is very typically Carignan on the nose, with peppery red fruit and an elegant austerity. Alain observed that good Carignan must be at least 13% abv. Le Mascarou, Côtes du Roussillon Villages Tautavel, is kept in old wood and is a blend of equal parts Carignan, Grenache and Syrah, with red fruit and good structure and length. La Carissa comes from Genegals, a vineyard in the Cirque de Vingrau and comprises 40 per cent each of Grenache and Syrah with 10 per cent each of Mourvèdre and Carignan. It was riper, with berry fruit, and more mouth filling, but with tannins and acidity.

Alain's Muscat de Rivesaltes comes only from Muscat à Petits Grains and is rounded and honeyed – Muscat works well in the limestone of Vingrau. A 2010 Rivesaltes Ambré was nutty and intense, and not quite *rancio*, while a Rivesaltes Tuilé, bottled as stock is needed, was dark tawny, with chocolate, prunes and liquorice. It was very concentrated with an incisive finish of acidity. L'Oublié was just that, a Rancio Sec, from a barrel that had been forgotten. It was severe with intense acidity and firm fruit, making a grand finale to a friendly meeting.

Domaine de l'Edre

Vingrau

www.edre.fr

Domaine de l'Edre is the result of a partnership between two friends, Pascal Dieunidou and Jacques Castany, who started the estate in 2001 with one red wine, producing just 3,500 bottles. They each had access to some vines, one through his wife, the other through his girlfriend, but had worked in other fields. In 2003 they made two wines, one oaked and one not, and gradually they bought more vines, in Vingrau, Tautavel and Opoul, and now have 10 hectares. In 2004 their wine scored well in a *Decanter* tasting, and in 2005 Robert Parker gave one of their wines a score of 95 and as Jacques put it, 'we haven't looked back'. They bought more vines and have developed a white cuvée. L'Edre is the *lieu-dit* of one of their vineyards, and means ivy in Catalan.

They have a small village cellar in the centre of Vingrau, where I met Jacques. Their white Côtes du Roussillon, Carrément Blanc is a blend of 50 per cent Grenache Blanc, 30 per cent Roussanne and 20 per cent Grenache Gris, half kept in wood, and half in vat, to yield some firm mineral notes and a tight-knit structure. Jacques observed that it was 'very calcaire', limestone being the dominant soil of Vingrau.

The first red wine was Carrément Mourvèdre, kept in vat, with a short *cuvaison* and a week's cool pre-fermentation maceration to retain the fruit. The red fruit was indeed very vibrant, with supple tannins. Jacques called it a summer wine. Carrément Rouge is a blend of 50 per cent Syrah, with 25 per cent Grenache, 20 per cent Carignan and 5 per cent Mourvèdre, kept in vat and given a three- to four-week maceration, making for some lovely ripe fruit. Their best red cuvée, L'Edre, is a slightly different mix of 60 per cent Syrah, 20 per cent Grenache, 15 per cent Carignan, and 5 per cent Mourvèdre from a selection of plots, each aged separately in barrel and assembled in the spring. A barrel sample was intense and rich, very chocolatey, and exuberant, with fresh tannin on the finish. 'That is Vingrau,' commented Jacques. I was then treated to a couple of more mature vintages; the 2016 was more restrained and the 2013 very rich and powerful, with silky tannins. It was very much the ripe style favoured by Robert Parker and his tasters, but successfully so.

Now that Jacques has given up his 'day job', he is clearly enjoying being fully involved with his wines. 'We make the wine we know how to sell,' he says. They have never made any Vin Doux Naturel.

TAUTAVEL

Domaine Fontanel

Tautavel

www.domainefontanel.fr

Domaine Fontanel is right in the centre of the village of Tautavel, just along the road from the cooperative, with cellars dating from 1864. The estate has recently changed hands. Pierre Fontanel, who first bottled wine here in 1989, sold the estate to Matthieu and Elodie Collet in 2017. Matthieu participated in the 2016 vintage and Pierre Fontanel is still around to offer advice and help. The original 35 hectares have been reduced to 20, with vineyards in Maury and Tautavel and near Calce and Baixas. Matthieu has done *stages* at Cazes and Piquemal. His parents were farmers in northern France and he has worked in drinks, specifically on filtration, in Calvados for cider, in Virginia, Bordeaux and Champagne. But when he realized that he really wanted to make wine he went to Montpellier to study. He had a *coup de coeur* for Roussillon – he talked of arriving in Calce at sunset – and three weeks later the SAFER (a government-run body which controls the sale of vineyards) had found him Domaine Fontanel. Matthieu's first solo vintage was 2018, a year that was complicated by mildew. He is starting to convert the vineyards to organic viticulture, entailing a financial investment of €150,000 over three years because of the need for different equipment, more personnel and smaller yields. Matthieu observed that he and his family feel very much at home here. He considers Maury more innovative, while Tautavel is more classic, and he likes the idea of old vines with their small yields. The village of Tautavel is dominated by the co-operative, which accounts for 80–90 per cent of the volume, with four or five *viticulteurs* accounting for half its production. Gérard Bertrand has a presence, but Tautavel can be vinified outside the village, which Matthieu considers will pose problems for the future. You sense that Matthieu is very committed to the region and its wines.

The range comprises a Côtes Catalanes Viognier, with some peachy fruit. A Côtes du Roussillon Rosé from Syrah and Grenache Gris, pressed directly without a malolactic fermentation, was rounded and vinous. Côtes du Roussillon Blanc comes from 40 per cent Roussanne, with the other 60 per cent a blend of Grenache Gris and Blanc, Macabeo, Carignan Blanc, Marsanne and Muscat d'Alexandrie. Half is vinified in new 300-litre barrels, and half in vat. It was quite solid and rounded.

Côtes du Roussillon Villages, from 70 per cent old Carignan, with Grenache and Syrah, was rich and rounded with black fruit. A red Côtes Catalanes, from approximately one-third each of Carignan, Syrah and Grenache marked a change in style from Pierre's wine. It was fresher, with a shorter maceration, ten days as oppose to three weeks, making for some elegant spice.

Some of the vineyards of Tautavel, with the appropriate schist soil, are included in the appellations of Maury and Maury Sec. Consequently, Matthieu makes Maury Sec, from 85 per cent Grenache with Mourvèdre. It was kept in vat and has fresh cherry fruit and an elegant backbone, a balance between fruit and concentration. Côtes du Roussillon Villages Tautavel, Cuvée des Cistes, an oak-aged blend of 60 per cent Syrah and 40 per cent Grenache, was quite firm and tannic.

As for Vin Doux Naturel, Matthieu began making Maury Grenat in 2017, with lovely ripe, spicy fruit. Muscat de Rivesaltes is nicely perfumed and best of all is Rivesaltes Ambré, a blend of Grenache Blanc and Macabeo, with intense, hazelnut fruit. The tasting finished with L'Ancestre Rancio Sec 2007, made by Pierre and bottled in 2018. It was amber in colour, very dry and intense with a firm finish. Matthieu and Elodie deserve to do well.

9

RIVESALTES

Rivesaltes is a small town just north of Perpignan, that gives its name to the largest appellation of Vin Doux Naturel in Roussillon. Another claim to fame is that it was the birthplace, in 1852, of First World War general Maréchal Joseph Joffre, who is commemorated with a statue in the main square. Joseph was the fifth of 11 children and his father, Gilles, was a vigneron and a cooper, or more precisely a *barricailleur*, who transported his own wine in his own barrels to a bourgeois clientele. Despite this claim to fame Rivesaltes itself is a rather sleepy town, where nothing much seems to happen; even finding a café on a Monday lunchtime proved quite challenging. Off the main street, there are some attractive narrow streets and a little square, where indeed there was a café. For finer dining, La Table d'Aimé, adjoining the cellars of Domaine Cazes is well worth a visit.

Outside the town are the vestiges of the camp that was first used to house refugees arriving from Spain after the Civil War, then gypsies and Jews waiting to be deported during the Second World War, and lastly, between 1962 and 1964, the Harkis arriving from Algeria. Pierre Boudau remembers that he had classmates at school who had been born in the camp. It was opened as a museum in 2015, a poignant reminder of the hardships of war, in the most desolate of spots.

The wines of Rivesaltes have a long history and have been appreciated over the centuries. Sadly, that is no longer the case today, as their reputation is for various reasons somewhat tarnished, even though the best examples are undoubtedly some of the finest Vins Doux Naturels of France. The earliest documentary evidence comes from 1394 when, as the Vatican archives record, Benedict XIII, the last of the Avignon popes, purchased through his *collecteur* in the diocese of Elne six charges of Vin Muscat de Claira. Claira is a village close to Rivesaltes. Roger Dion discusses this evidence in his authoritative history of French wines, stating that evidently no wine of this type was made in the Rhône Valley at that time and that hitherto this style of wine had come from Greece, Crete or, best of all, Cyprus. During the fourteenth century, after Arnaud de Villeneuve's discovery, Grenache rather than Muscat assumed prominence as the grape variety for *vin de liqueur* or fortified wine. The Spanish influence is very much in evidence here, for Grenache (or Garnacha) is Spanish in origin and Roussillon was under the rule of the kings of Aragon.

In the eighteenth century Voltaire sang the praises of Rivesaltes. He was the uncle of Monsieur de la Houlière, governor of the fortress of Salses, the magnificent fortification that still stands at a strategic point on the road between Perpignan and Narbonne. Voltaire wrote that he 'experienced great pleasure when [he] drank a cup of the Salses wine,' although his 'frail human mechanism was unworthy of that liquor.' When travelling through France in 1787 and 1788, Arthur Young also enthused about Muscat de Rivesaltes. Rostand referred to it in Cyrano de Bergerac, while Grimaud de la Reynière, writing his Journal des Gourmands et des Belles in the early nineteenth century considered it the best fortified wine of the kingdom.

All the nineteenth-century authorities rated Rivesaltes highly. Muscat seemed to have superseded Grenache as the principal grape variety, for in his *Topographie des Tous les Vignobles Connus* of 1866 André Jullien described it as the best Muscat, not just of France, but of the whole universe, when it came from a good vintage and had aged for ten years. James Busby visited Rivesaltes in 1833, during a journey to France and Spain to collect vine cuttings to take back to Australia, and described how the grapes were left to shrivel on the vines. Jules Guyot rated the Muscats of Rivesaltes more highly than those of the Hérault, such as Frontignan and Mireval, describing how the wines were made with dried grapes with no addition of spirit. Cavoleau also enthused about the Muscat from Rivesaltes, explaining that there were three grape varieties, Muscat Alexandrin, Rond Blanc and best of all St Jacques, that were harvested in two pickings. The first grapes were left to dry until the others had been picked.

In 1857 Victor Rendu gave the most detailed appraisal in his *Ampélographie Française*, rating Rivesaltes as the most important wine of the department of the Pyrénées-Orientales, covering 10,500 hectares, of which half were on slopes. The grape varieties included Carignan, Grenache, Mataro, Picpoul Noir and Clairette. Carignan was planted in three-quarters of the vineyards and blended with Mataro (Mourvèdre), which gave body and colour, while Grenache contributed sweetness and vivacity. What he called *vin de commerce* formed the principal income of Rivesaltes, while Salses and Baixas were also mentioned as other important parts of the vineyards. The wines with a universal reputation were Muscat, Macabeo, Malvoisie, Grenache and Rancio. Rendu went on to explain how the Muscat was dried on the vine and in the first year was more like a syrup than a wine, and how the vinification for Macabeo and Malvoisie was slightly different as they were not dried. He also mentioned the wines of Perpignan, including some excellent *rancios*.

In 1872 the so-called Arago Law recognized the existence of a style of winemaking that was peculiar to Roussillon and in 1898 the term Vin Doux Naturel was given legal recognition. The impact of phylloxera was catastrophic, and although the vineyard was subsequently reconstituted, it never regained its former size. Nevertheless, fortified wines were popular enough before the Second World War to make Rivesaltes one of the earliest appellations, in 1936, along with Banyuls and Maury, and also Côtes d'Agly and Côtes du Haut-Roussillon, which were subsequently incorporated into Rivesaltes in 1972. A lesser appellation, Grand Roussillon, was created in 1938 as a secondary appellation for wine that was not of the quality of Rivesaltes, Maury or Banyuls, but that is now defunct. Muscat de Rivesaltes, including any Muscat grown in Banyuls and Maury, as well as Rivesaltes, and covering 90 villages in the Pyrénées-Orientales and nine in the Aude became an appellation in 1956. Any Muscat produced on the Côte Vermeille is also Muscat de Rivesaltes, rather than Banyuls.

The appellation of Rivesaltes covers all the vineyards of Roussillon with the exception of the Côte Vermeille. Senator Gaston Pams, an important political figure in the department in the 1920s and a keen defender of viticulture in the region, argued that the vineyards on flatter land should also profit from Vin Doux Naturel, so that the appellation of Rivesaltes was extended all over the region, in an attempt to help vineyards with some quality issues. Rivesaltes was chosen as the name for the appellation because the town was at the heart of the négociant activity of the region and an important railway junction, and a hub for wine transported through the department. It is at the end of the line of what is now a tourist train, Le Train Rouge, which was originally used for transporting wine down the Agly Valley.

As discussed on page 20, the appellation of Rivesaltes covers various categories and colours. The following profiles cover those based in Rivesaltes itself and the neighbouring village of Baixas, as well as the estates outside Perpignan, in an attempt to demonstrate some of the diversity of the extensive appellation.

Cave Arnaud de Villeneuve

Rivesaltes

www.arnauddevilleneuve.com

The cooperative that dominates the production of Rivesaltes has its ageing cellars and a smart shop front in the main square of the town. The statue of Maréchal Joffre is further along the square. I admired the extensive barrel cellar; the oldest wine dates from 1955 and they also have a few vintages from the 1960s.

These days the cooperative is a fusion of several other cellars. In Rivesaltes itself there were originally three cooperatives, and subsequently they have joined up with those of Pézilla-la-Rivière and Corneilla-la-Rivière and then more recently, Salses. The director, Jean-Pierre Papy, gave me an extensive tasting and talked about the cooperative's work and ambitions. He has a broader view of things, having worked in communications and consultancy in Paris before coming south, first to Cave Abbé Rous, the large cooperative in Banyuls, and then to Rivesaltes, where he has been since 2010. The cooperative currently includes 246 members, with a total of 1,900 hectares, and 48 of those members accounting for 50 per cent of the production. The cooperative was a pioneer, the first to plant Chardonnay in Roussillon in the early 1980s. It has seven members who farm organically, accounting for 125 hectares, with an eighth in the process of conversion. They are working on joining the association of Vignerons en Développement Durable, an organization which concentrates on environmental awareness, taking into account social and economic considerations. Its members have not used any weedkiller since 2017.

Jean-Pierre explained that Muscat is complicated and not really appreciated. Their Fresh Muscat was just that, with an emphasis on the fruit, and relatively low alcohol, sold in a modern package. They also make a *moelleux*, with 40 grams per litre of residual sugar, like a Muscat de Rivesaltes but without the added alcohol. A Côtes du Roussillon Blanc is a blend of Macabeo and Grenache Blanc, with a hint of Vermentino and Roussanne, with some herbal fruit. They are developing a range of wines under the name No.153, after the address of an old cellar, with a greater emphasis on particular plots and wine growers. They have bought some polymer eggs for the Grenache Blanc. There is a range of Côtes du Roussillon, based on Grenache and Syrah. A Côtes du Roussillon Villages called RD 900, from a selection of vineyards in Pézilla-la-Rivière and Espira-de-l'Agly, was refreshingly unoaked with ripe fruit. Oppulum Côtes du Roussillon Villages comes from a distinct area near the village of Opoul, a plateau at 250 metres altitude, where the grapes ripen ten days later than on the plain. The cooperative in Opoul closed so the wine growers there had to choose between Tautavel or Salses. The Syrah and Grenache are picked at the same time and the élevage takes place partly in new 500-litre barrels and partly in amphorae. The wine was structured, with spicy fruit.

We moved on to Vin Doux Naturel and Jean-Pierre talked dispassionately about the problems of selling Vin Doux Naturel. 'You cannot deny that the market for Rivesaltes is declining,' he says. 'The export market is particularly difficult. We need to teach the younger generation about these wines; they must not be forgotten, but there is no miracle solution.' Muscat de Rivesaltes was rich and honeyed with notes of orange. It was muted at 115 grams per litre (the range is 100–120 grams per litre) and the volume of alcohol is 10% abv. Originally the appellation required the use of both Muscat d'Alexandrie and Muscat à Petits Grains, but that is no longer the case. A 20-year-old Ambré Hors d'Age is a blend of Macabeo, Grenache Blanc and Grenache Gris, with rounded, nutty fruit and a firm bite on the finish. Then came some vintage wines. The 1988, a blend of Macabeo and Grenache with just a touch of Muscat, was honeyed and rich, with *fruits confits*. In contrast, 1979 was dry and nutty, with hints of prunes. They use barrels and *bonbonnes* for ageing these wines; the *bonbonnes* are outside and the wine spends no more than two years in them. Our tasting finished with the 1970, which reminded me of an old oloroso; it was austere but rich and so long that I could still taste it when I arrived for my next appointment. These wines are ridiculously cheap for their quality, €69 for a wine that is 50 years old. They must be saved.

Domaine Boudau

Rivesaltes

www.domaineboudau.fr

I tracked down Pierre Boudau in a narrow back street of Rivesaltes. His grandfather started the family business, but his father died quite young, so it was his mother who continued his grandfather's work. The wine, Vin Doux Naturel, was all sold in bulk, but he and his sister, Véronique, decided to set up as an independent cellar in 1992. They have 55 hectares, in two main areas, on the plateau of Baixas with *galets roulés*, and near Vingrau with vines on a rocky limestone and clay hillside. Pierre talked about the differences between the two areas. The season in Baixas, with its maritime influence, is 15 days ahead of Vingrau, emphasizing the sucrosity of the Grenache, whereas in Vingrau you obtain much more finesse. They are now converting to organic viticulture. It is Véronique who makes the wine, learning from their oenologist, rather than any formal studies.

Our tasting began with a dry Muscat, with a classic vinification. It was rounded and grapey, with a slightly bitter finish. Pierre observed that dry Muscat is difficult to sell and yet it goes very well with asparagus and Asian cuisine. Sweet Muscat is even more difficult to sell. Côtes du Roussillon Le Clos comes mainly from Vermentino, with some *bâtonnage* and ageing on the fine lees, making for elegant weight and herbal fruit. Pierre considers Vermentino to work well in Roussillon, where it is quite productive and retains its freshness. Côtes du Roussillon Blanc, Henri Boudau, after his father, is a blend of Grenache Blanc and Macabeo, given four months in barrel. It was rounded and floral, with a touch of fennel and some integrated oak. Pierre explained that he has gradually increased his range, from a small beginning of Muscat and rosé. His Côtes du Roussillon Rosé has rounded, red fruit and a vinous palate, coming mainly from Grenache.

There is a logical range of red wines. Le Petit Closi, a Côtes Catalanes from Grenache, Syrah, Carignan and Cabernet Sauvignon, was rounded and supple. Pierre explained that he had wanted to call it Petit Clos, but Clos is not a permitted term for an IGP. Côtes du Roussillon le Clos, mainly Grenache with some Syrah, from the Baixas plateau, is nicely spicy. Next came Tradition Côtes du Roussillon Villages, from Syrah, Grenache and Carignan, and then Patrimoine Côtes du Roussillon Villages, made from Grenache with some Syrah, from Vingrau, which was the more structured of the pair, with a fresh finish. Côtes du Roussillon Villages Cuvée Henri Boudau was mainly Syrah with some Grenache, and deep in colour with oak and structure. Pierre talked about the subtle changes in their winemaking over the years, picking a little earlier these days and aiming for less ripeness so that the wines are lighter, with less extraction. They do gentle *pigeage* and *mouillage*, or dampening of the cap. He feels that Roussillon is still a young vineyard and that they are in the phase of research, as yet nothing is fixed.

Our tasting of Vin Doux Naturel began with a Rivesaltes Grenat, with ripe, red fruit. The best Grenat comes from old vines, of which they are lucky to have six hectares, which were grafted in 1967, with the American rootstock planted the previous year. Pierre's father made a special journey to Châteauneuf-du-Pape to select his vines, his *greffons*. A Tuilé, *muté sur jus*, which is a less complicated process than *mutage sur grains*, was ripe and nutty. We finished with an opulently sweet Muscat de Rivesaltes from an almost pure Muscat d'Alexandrie. It was ripe and intense with acidity on the finish. Pierre does not consider Muscat à Petits Grains to be suitable for the terroir and thinks it is better for dry wine.

Pierre talked about the problems of running a business from the centre of town, saying that, 'The mayor wants us out, and also Domaine Cazes.' They have built a new vinification cellar, used for the first time in 2019, on the outskirts of Rivesaltes. He also talked about the problems besetting Rivesaltes, giving me plenty of food for thought about the difficulties of the appellation, and of Vin Doux Naturel in general.

Domaine Cazes

Rivesaltes

www.cazes-rivesaltes.com

Domaine Cazes is a family company that is now part of the extensive group AdVini. However, the winemaking policy is still very much in the hands of the family. Michel Cazes was a market gardener at the beginning of the twentieth century, who grew vines as well as vegetables, and made some wine for the local market. His son, Aimé, decided to expand the family vineyards, an endeavour that was continued by his two sons, André and Bernard. Bernard's son, Emmanuel, is now the public face of Domaine Cazes, their wine ambassador. He used to be the technical director, but now Isabelle Roi makes their wine in Rivesaltes. Lionel Lavail is the managing director. As well as an extensive 180 hectares of vines close to the town of Rivesaltes, they bought Clos des Paulilles in Banyuls in 2013, and have other plots of vines in Roussillon, a total of about 330 hectares, making them the largest private vineyard owners in Roussillon. Domaine du Grand Chêne in Les Aspres, with 30 hectares, is a recent purchase, which Emmanuel described as 'a clean sheet' with which they will be able to experiment, working on wines with lower alcohol and no sulphur. Cazes see themselves as pioneers for organic viticulture in Roussillon and altogether 250 hectares are farmed organically. They began ploughing their vineyards in the 1990s and after some initial experiments in 1997, inspired by the problems of flavescence dorée and oidium, for which conventional chemical treatments did not work, started the conversion to organic viticulture in 1998.

I spent a convivial hour with Emmanuel tasting as he updated me on Domaine Cazes since my previous visit. Canon du Maréchal was a pioneering brand and accounts for half their production, in all three colours. The rosé is a blend of Syrah, Mourvèdre, Grenache and Carignan, with just a hint of Muscat. Canon Blanc is based on Muscat, with some Viognier, with pithy fruit and fresh fruit. Emmanuel observed that people in Roussillon do not drink Muscat Sec whereas elsewhere there has been a move to more aromatic varieties. Canon Rouge is a blend of Syrah and Grenache, which marks a distinct shift from ten years ago when the blend was Cabernet and Merlot. I remembered my first taste of Canon Rouge, with André Cazes sometime at the end of the 1980s, when it included Syrah, Merlot, Cabernet Sauvignon, Grenache Noir, Cinsault and Carignan, making a soft, fruity glass of wine. These days it has fresh spice with a supple finish.

Côtes du Roussillon, John Wine, a play on words of John Wayne, is a blend of Syrah and Grenache, with fresh spice. Ego, Côtes du Roussillon Villages, is at least 50 per cent Grenache with some Syrah and a small amount of Mourvèdre, with an élevage in vat making for ripe fruit with silky tannins. A second Côtes du Roussillon Villages, Alter, is a blend of 60 per cent Syrah with Grenache and some Mourvèdre, for which they use some interstaves, so there was vanilla on the nose, with ripe fruit and a youthful streak of tannin. Le Credo, Côtes du Roussillon Villages has also changed to southern varieties. The first vintage, 1993, was a blend of Cabernet Sauvignon and Merlot and remained so until 2005. None was made in 2006 and then in 2007 it became Syrah and Grenache with some Mourvèdre and Carignan, with an élevage in concrete eggs and small foudres. The vines are at Montner at 160 metres. There was fresh, youthful fruit and an elegant concentration. Emmanuel enthused about Carignan, pointing out that it retains acidity, balancing the very ripe Grenache.

Then we moved on to Vin Doux Naturel, beginning with a 2004 Rivesaltes Ambré, which was soft and honeyed, with notes of brioche and *pain d'épice*. It comes from Grenache Blanc, aged for 11 years in foudres. Since 2012 they have used organic alcohol for the mutage; it comes from organic grape skins and costs significantly more than normal alcohol, at €200 per hectolitre rather than €90–€100. You cannot taste the difference, but as organic producers, they must respect the authenticity. Emmanuel observed that at one time his father sold more Vin Doux Naturel than the whole AdVini marketing team today. He feels strongly there is a place in the market for the best Vin Doux Naturel. Rivesaltes Grenat, from Grenache Noir, has ripe, round, spicy fruit. For Emmanuel, there is a difference between this and Rimage from Banyuls, with Rivesaltes making less Grenat than Banyuls makes Rimage. A Tuilé from Grenache Noir, aged for seven years in foudres, was rounded, with red fruit, walnuts and prunes, and a streak of tannin.

Emmanuel explained that the grapes for dry Muscat need to be picked at the end of July; they use Muscat d'Alexandrie, and prefer Muscat à Petits Grains for the Muscat de Rivesaltes, with about 30 per cent Muscat d'Alexandrie. The sweetness levels have changed, with a trend towards less sugar. Twenty years ago, they muted at 120–125 grams of sugar per litre, as opposed to 105–110 grams of sugar per litre these days, with more alcohol to compensate. Our tasting finished with 1978 Cuvée Aimé, an Ambré Hors d'Age, from 70 per cent Grenache Noir and 30 per cent Grenache Blanc, first bottled in 2000. It is named after Emmanuel's grandfather, Aimé, who died in December 2000, at the grand age of 99, and it has for some time been one of my favourite Rivesaltes, with its elegant, nutty fruit and long, lingering flavours, that are impossible to capture in words.

Dom Brial

Baixas

www.dom-brial.com

The cave cooperative of Baixas is an important producer of Rivesaltes as well as vin sec. Today it has 200 members, with 2,100 hectares of vines, and accounts for 10 per cent of the total production of the Pyrénées-Orientales. White and rosé wine together, red wine and Vin Doux Naturel each account for one-third of their production. With the highly successful brand Dom Brial, named after an eighteenth-century Benedictine monk who left money to found a school in the village of Baixas, they are the largest producer of Muscat de Rivesaltes. The cooperative was officially set up in 1923, but its history goes back to the wine growers' demonstrations of 1907, with some of those protestors going on to create a first cooperative. The buildings are next to the old station at Baixas, from which the line ran to Rivesaltes.

Xavier Ponset, who runs the welcoming *caveau de vente*, took me to see the vineyards of Crest Petit, on a hill above the village – Baixas means *bas*, or low, with the village in the valley bottom. The hills are the first foothills of the Corbières and still have the old stone walls that were used by shepherds as sheep pens, and also the remains of an old quarry. The blue marble of Baixas was used in the building of the cathedral in Washington DC. There were dramatic views of Força Réal and the Canigou in the distance as we looked down over Baixas, with its church tower dominating the russet roofs of the village. The spectacular altarpiece in the church is worth a trip in itself, with its frieze of vine leaves and grapes. In the eighteenth century the vineyards were on the plains but they were moved up into the hills in the nineteenth century. In 2005 the cooperative bought some *garrigue*-covered hillsides which they cleared in order to plant Syrah and to experiment with organic viticulture. They have ten members in conversion.

The cellars contain numerous small vats, enabling them to work *à l'hectare*, with multiple small selections. There are at least 2,000 barrels, two-thirds for Vin Doux Naturel and one-third for red wine, as well as large foudres for Vin Doux Naturel. Baixas has never used *bonbonnes*, only barrels, for ageing its Rivesaltes. Château les Pins, a property in Baixas, is the flagship vin sec of the cooperative, bought in the 1980s, with a first vintage in 1991. The cooperative is part of Vignerons Catalans, but not one of the original members, joining after the crisis of Vin Doux Naturel in the mid-1990s. However, for Xavier, their own Vin Doux Naturel represents their identity and their wealth.

First we tasted Crest Petit, a vin sec from Syrah, with some Grenache, planted on lower hillsides and given 12 months of élevage in demi-muids. The rich fruit on the palate was balanced with freshness and silky tannins. Next came some Muscat de Rivesaltes. Xavier explained that while the Cuvée Dom Brial favours the classic youthful fresh fruit, their Muscat from Château les Pins always comes from an older vintage. The 1994, when it was 15 years old, was extraordinary, amber golden in colour, with a nose of mature Sauternes, honeyed bitter-orange notes, a balance of sugar and acidity and a smooth, fresh finish.

Next came a vertical tasting of Rivesaltes Hors d'Age, all based on Grenache, and possibly some Macabeo, and bottled in 2009. The 1989 was redolent of walnuts and hazelnuts with some dried fruit on the palate and a long, intense finish. There was more acidity in the 1979, and notes of *rancio,* with a firm, nutty palate. The 1969 reminded me of old oloroso: all the sugar had gone and it was firm and concentrated with a fresh finish. The 1959 looked like Madeira with a yellow-green rim, and was intense and nutty, with a more incisive finish. This is the oldest wine that they still have in barrel. I felt I was tasting the history of Roussillon. Our tasting finished with a 2000 Rancio Sec bottled in 2017, made from Grenache Gris and Grenache Blanc, with a dry nose and a salty, austere palate. It was like a fino sherry, but somewhat heavier, and quite delicious.

Clos des Fées

Rivesaltes
www.closdesfees.com

Hervé Bizeul has the dispassionate objectivity of an outsider. He worked as a journalist in Paris, writing about wine before he learnt how to make it. You sense that he is a free spirit and that he does not really care what people think about his wines. 'I do not have a lot of ambition for my wines. I accept what nature gives me, and I enjoy myself. I am free,' he says. He has certainly broken down some barriers. His first serious vintage was 2004.

He explained that he has four different terroirs; the schists of Cases-de-Pène, the schists of Maury, the granite of Lesquerde and the limestone of Calce. Essentially, he makes three styles of wine, what he called *vins de fruit,* such as a white Côtes du Roussillon from Vermentino, with some Grenache Blanc and Roussanne. We shared an enthusiasm for the white wines of Corsica, from Vermentino. Then there are *vins de lieu* with a sense of place, made from old vines, and finally wines from grape varieties that have never been planted here before, such as Pinot Noir, Cabernet Franc, Tempranillo and Semillon. Altogether he has 50 hectares in some 130 different plots. Hervé used to make his wines in an old village cellar in Vingrau, but 2019 saw him complete his third vintage in a streamlined building on the edge of Rivesaltes. There are lots of small stainless steel vats as well as tapered stainless steel vats, with wider tops than bottoms, which Hervé explained allow for a gentler extraction.

Hervé has a quick mind, even when he is jet-lagged, having returned from Singapore less than 24 hours earlier, as was the case when we met.

He enthusiastically opened numerous bottles, and several wines stood out. The most original wine is undoubtedly Un faune avec son fifre sous les oliviers sauvages, from 50 ares of Muscat onto which he has grafted Semillon. It was a wonderfully original expression of that variety, with a golden colour and some rounded, waxy fruit, balanced with acidity, length and texture. Les Sorcières is a blend of Syrah, with some Grenache and Carignan, as well as, unusually, a drop of Cinsault. It has elegant, fresh fruit and supple tannins. Hervé observed that wine is like cooking – one should not do too much. He quoted Nicolas Joly, the pioneering biodynamic producer from Savennières, who said, 'before being good, a wine must be true to itself.'

His Côtes du Roussillon Villages comes from Grenache with some old Carignan and a little Syrah, and is fermented and stored in a large concrete vat. He considers it very typical of the appellation, with firm fruit, good structure and length, combined with elegant concentration. Hervé could not resist enthusing about his own Le Clos des Fées; it proves that you can make great wine in the south. It is made with meticulous care, from Syrah planted at a high density of 8,000 plants per hectare, plus some Carignan, Mourvèdre and Grenache Noir, blended before an 18-month élevage in new wood. The wood helps the structure of the wine, enabling it to age; the oak was very present in the young wine, but there was fruit underneath. It is a wine that needs several years of bottle ageing, with Hervé adding that you need financial security to be able to stock wine.

La Petite Sibérie made its reputation as one of the most expensive wines of the region: Hervé was selling the 2013 vintage in October 2019 for €200. It comes from a 1.16-hectare plot at Cases-de-Pène. Grenache is topped up with Syrah and Mourvèdre, to respect the appellation rules, and is given some élevage in new oak. When it was seven years old, the wine had ripe liqueur cherry fruit, the benchmark of Grenache, with a firm grip of tannin. Whether it is worth €200 is another matter. In complete contrast, the next wine was Modeste, which retails at €8.50 a bottle. Young vines and underripe grapes make for refreshing berry fruit, what Hervé called 'tutti frutti'. His Pinot Noir comes from a 1.7-hectare north-facing clay and limestone plot between Vingrau and Tautavel, planted in 2012. The vines were a clonal selection from Clos de Tart and the wine was very silky, with elegant, red fruit and a tannic streak. There was also a Tempranillo, Images Dérisoires, with rounded spice and a hint of liquorice and Un Faune, with rounded cassis fruit and good length, which comes from Cabernet Franc, with some Cabernet Sauvignon and Merlot: 'Approximately the blend of Château Cheval Blanc,' observed Hervé. De battre mon coeur s'est arrêté is a pure Syrah with fresh peppery fruit, from vines in Vingrau and Tautavel, and also Lesquerde, where the granite is particularly good for Syrah.

My tasting finished with the 2016 Passat Minor, a Vin de France Liquoreux made from *passerillé* grapes. It had 12.5% abv with 200–260 grams per litre of sugar. The colour was old gold and the nose perfumed with hints of orange, while on the palate it was very intense, with honeyed and *confit* notes, and acidity on the finish.

The flatter land near the coast, almost in the suburbs of Perpignan and towards Canet-en-Roussillon is not necessarily the most suitable land for vineyards, but none the less there are a handful of noteworthy estates.

Mas Baux

Canet-en-Roussillon

www.mas-baux.com

Mas Baux is a small estate with vines on *galets roulés* like those of nearby Château de Rey (see below). Serge Baux explained that the *lieu-dit* of the estate is in fact Mas Durand. The Durand family were here in the 1700s and at the time were important wine merchants, and one of the family was a *deputé f*or the Hérault. Another member of the family supplied Napoleon's troops with wine, when they were invading Spain. The large property was split up and in 1997 Serge bought a house and 12 hectares of vines from a 92-year-old spinster called Mademoiselle Durand.

When Serge arrived here following a career in the glass and ceramic industry the property needed extensive renovation. He pulled up a lot of vines, replanting for vin sec, rather than Vin Doux Naturel, but did keep two hectares of old Grenache. He bottled his first wine in 2002. Unusually he planted Cabernet Sauvignon, aspiring to imitate a rosé from Penedès that he had particularly enjoyed. The vineyards are about three kilometres from the sea, on a terrace at just 40 metres altitude. Serge explained how the size of the *galets roulés* can vary, depending on how far they have come from the mountains. Here they have come from the Pyrenees, which are close – you can see the Canigou in the distance – consequently the *galets* can be very large. In contrast, the *galets* of Costières de Nîmes in the southern Rhône come from the Massif

Central and are smaller, while the *galets* of Châteauneuf-du-Pape come from the Alps, and have travelled even further, and consequently are even smaller. The vineyards are all farmed organically; Serge uses pheremonic controls to combat *vers de la grappe* and part of his vineyard is a trial for a Spanish laboratory that has created a pheromone control for a new pest (see page 44).

The cellar building is a 300-year-old barn. At the end of the nineteenth century silk worms were produced here. These days it is well equipped with stainless steel vats, with a system of mechanical *pigeage* that dampens the grape skins. Everything works by gravity. The names of Serge's wines are quite colourful. Autant en emporte le Blanc – a play on Autant en emporte le Vent, or Gone with the Wind – is a blend of Grenache Gris and Grenache Blanc, fermented and aged in barrel, with tight-knit acidity. Rouge à Lèvres (lipstick) is a rounded rosé, including the Cabernet Sauvignon along with Grenache and Mourvèdre. La Vie en Rouge (life in red, as opposed to pink) is a fresh, peppery Côtes Catalanes Syrah and Velours Rouge (red velvet) is a Côtes du Roussillon blend of Grenache and Syrah, with spicy, red fruit. Serge loves Mourvèdre and uses it to make Mataro Boy, a Côtes Catalanes aged in vat, with fruit and structure. Soleil Rouge is Mourvèdre with some Syrah, given a year's élevage in wood, with some firm, peppery spice. My tasting finished with the very first wine he made, Grains de Beauté (beauty spots), a Rivesaltes Tuilé bottled in 2005, after five years in barrel. It was rich and intense, with notes of walnuts and chocolate.

Domaine Lafage

Perpignan (Canet-en-Roussillon)

www.domaine-lafage.com

Domaine Lafage is based at Mas Miraflors, outside Canet-en-Roussillon. Altogether the business includes three estates, the largest near the sea, as well as 20 hectares in Les Aspres, and the estate of St Roch in Maury, which they bought in 2007, with its own cellar. In addition, there is a recent purchase of 17 hectares in St Paul-de-Fenouillet for white wine, and another acquisition in Rivesaltes is being converted to organic viticulture, making a total of 260 hectares. Their beginnings, however, were small. Jean-Marc and Eliane's first vintage was 1995, when they made just 1,000 bottles of a Muscat de Rivesaltes. Then, in 2001, Jean-Marc took over the family vineyards and acquired Mas Miraflors in 2006. The Lafage family came from Maury; Jean-Marc's grandfather was one of four brothers, three of whom settled in Canet, while one stayed in Maury at Domaine du Dernier Bastion.

When I visited in September 2019 the harvest was in full swing, so I was only allowed to see the cellar from a walkway above. The harvest began with Muscat on 6 August and was planned to finish on 10 October. We watched the antics involving an OVI, a steel vat that is taken into the vineyard on the back of a truck. The grapes are sealed in it without any air – Jean Marc is firmly anti-oxygen. The OVI was being manoeuvred by a pulley into position above a vat, into which the grapes would be released. It looked a little how I imagine a docking space ship might. Jean-Marc and Eliane have both worked in cellars in other parts of the world and their own cellar is inspired by the best that they have seen elsewhere. There is a forest of stainless steel vats, with barrels of different sizes and from different coopers. As well as both being qualified winemakers, they also employ five oenologists. The scale is pretty breathtaking.

The business is certified for HVE level 3, and they are working on organic viticulture, with two weather stations in the vineyards to help them plan their treatments. Jean-Marc observed that the market is increasingly asking for organic wines. In the cellar, it is essential to respect the fruit and they would both say that the key to their wines is freshness and elegance with a true expression of the terroir. Jean-Marc enthused: 'Here in Roussillon we have a dream terroir; travel makes us appreciate what we have here. Carignan is a little treasure; Maury is where they discovered Carignan.' It is the red variety with the most acidity, with as much as in white grapes, and provides a backbone for the wines.

I was treated to an extensive tasting, of which the highlights were several. White Côté Est, Côté Rosé and red Côté Sud provide easy drinking for their respective colour. Miraflors Rosé was nicely rounded. White Cuvée Centenaire, including some 100-year-old Grenache Gris and Grenache Blanc as well as Roussanne, was elegantly textured, with a fresh finish. La Grande Cuvée Blanc from Grenache Gris, Grenache Blanc and Macabeo, aged in foudres and barrel, has well-integrated oak and good acidity. Taronja is a characterful orange wine, made with Justin Howard-Sneyd from Domaine of the Bee and is a recent introduction to the range (see page 55).

As for red wines, Authentique is a blend of Syrah, Grenache and Carignan from different terroirs that represent Roussillon, with spicy fruit and tannin. Léa, named after their daughter, comes from Les Aspres and is fresh and ripe, with structure. Narassa is another recent addition, beginning as a mistake when some grapes were left forgotten in the vineyard and picked later than usual. You could almost equate it to a French Amarone, with peppery but ripe fruit and an intriguing finish. Le Vignon from Carignan, Mourvèdre and Syrah is nicely concentrated, with tight-knit fruit.

Naturally my tasting finished with Vin Doux Naturel, a fresh Muscat de Rivesaltes, with its grapiness exploding in the mouth. An Hors d'Age Rivesaltes, from a solera begun in 1998, was nutty and rounded and a Grenat Maury, mainly from Grenache Noir, was fresh, with ripe red fruit. Both Jean-Marc and Eliane exude energy as well as enthusiasm and determination. I left with a feeling that Domaine Lafage has come far, and will continue to go far, establishing its position as one of the more successful players in Roussillon.

Château de Rey

Canet-en-Roussillon
www.chateauderey.com

Château de Rey is located on the plain between Perpignan and Canet-en-Roussillon. Although it is surrounded by vineyards and fields, you sense that urban life is not that far away. Except for a brief period when it was rented, the property has been in Philippe Sisqueille's family since 1875, when his grandfather, who was already a vine grower and négociant, bought it from a M. Rey, a local notable who built the château but then went bankrupt. In 1996 Philippe bought back 40 hectares of vines with the cellars, but not the château itself. I admired the enormous foudres – the largest of which is 425 hectolitres – but they are no longer used, as they once were, for Vin Doux Naturel.

Cathy Sisqueille, Philippe's wife, explained that they are only interested in the Mediterranean varieties; Grenache Gris, Grenache Blanc, Macabeo, Roussanne, Vermentino and the two Muscats for white wine, and, for red wine, Grenache Noir, Carignan, Syrah and Mourvèdre. Their vineyards are all in one large plot, and no more than 1.5 kilometres from the sea as the crow flies, so the vines benefit from the fresh sea breezes.

All their wines are Côtes du Roussillon except their entry level range, Oh de, which is IGP Côtes Catalanes. A level up from Oh de is Sisquo, then Galets Roulés, from selected plots, finishing with Mine de Rey and a slightly sweet Muscat d'Alexandrie, simply named C29, after the vat number. Mine de Rey is a play on words, *mine de rien* or, it doesn't seem like it. The white is a blend of Grenache Gris and Grenache Blanc, grown on *galets roulés*, and fermented in 400-litre barrels made, unusually, from acacia as well as oak. Although there is no tradition for acacia in Roussillon, they had wanted to try it and followed their cooper's advice of mixing the two woods in the same barrel. Cathy likes the freshness and tension that acacia gives the wine. Galets Roulés Rosé is unusual for its fermentation and ageing in barrel, which gives the wine more weight. The red Galets Roulés has pleasing concentration, without being too heavy. Cathy observed that they had shortened the length of their macerations to make the wines more drinkable. Red Mine de Rey is a blend of 80 per cent Mourvèdre with some Grenache and Syrah, aged in wood to produce some firm tannins and ripe fruit.

My tasting finished with Vins Doux Naturels of which the highlights were 2005 Vent d'Anges, a pure Muscat d'Alexandrie, aged for two years in barrel and three in vat, so that it was bottled in 2010. It was rich and honeyed, prompting Cathy to observe that these are wines to philosophize about, not to sell. The 2007 Le Chouchou Rivesaltes Tuilé, a pure Grenache Noir bottled in 2017, has wonderful elegantly nutty fruit and a long finish. Cathy, who came to wine, as she put it, 'for the love of it', enthused that no vintage is ever the same.

Domaine Rière Cadène

Perpignan

www.rierecadene.com

Domaine Rière Cadène, which takes its name from owner Jean-François' mother's maiden name, is an oasis of tranquillity just beyond the outskirts of Perpignan. You leave a fast dual carriageway and find yourself rather unexpectedly in fields and vineyards. As Jean-François Rière observed, Perpignan has grown rapidly, but they are defending their corner, not just for the vineyards, which are organic, but also for the hoopoes that feature on their labels. Altogether he and his wife Laurence, and now their nephew, Guillem Batlle, have 30 hectares of vines. As well as those outside Perpignan they also have vineyards further inland in the Agly Valley on schist and black marls at Espira-de-l'Agly for red wines, and at Calce and Baixas, on limestone, for white wines. Jean-François enjoys having vineyards in three different places because he can juggle the terroirs. He is particularly enthusiastic about the terroir of Calce, with its old Carignan and old Grenache. Although Calce seems a world away from the outskirts of Perpignan, he assured me that it is only 25 minutes by tractor.

Their cellar is outside Perpignan for historical reasons. Jean François' great-grandfather had his cellar there, and sold wine in bulk. His grandfather was a prisoner of war, and on his return joined a cooperative that no longer exists. Jean-François remembers that his grandfather used a horse for the harvest as recently as the 1960s. His father worked in the car industry, but kept his vines, and then Jean-François and Laurence took over the estate 25 years ago, bottling their first wine with the 1996 vintage. Guillem has brought new vision to their sales. In three years, they have trebled their production in bottle, and nowadays 90 per cent of sales are vin sec, an exact reversal of the situation 25 years ago.

I tasted a selection of wines. For vins secs, J'ai Rendez-Vous avec Vous, Côtes Catalanes is the entry level in three colours, then comes Huppée. This is followed by Augusta, Côtes du Roussillon, in red and white, named after the Roman road to Barcelona, a continuation of the Via Domitia. The range is rounded off by a selection from individual plots. As well as Macabeo and Grenache Blanc, Jean-François has Vermentino, Roussanne and Viognier. He is particularly enthusiastic about the potential for white wines in the region and wants to expand his production. Reds tasted included Augusta, from Grenache and Syrah, which is redolent of fresh, easy spice. A pure Syrah, Les Galets de la Têt, Cuvée Jean Rière, was a blend of 70 per cent Grenache with 30 per cent Syrah, with spice and concentration. Fernand Cadène, named after Jean-François' grandfather, is a blend of Syrah and Grenache.

Much as I enjoyed tasting Jean-François's vins secs, it was his Oxy range of oxidative wines that really impressed me. No 1 is Macabeo which had developed a *voile*, like fino sherry or vin jaune. It was bottled after eight years in wood and was firm and intense, dry and nutty with a salty finish. No 2 is a Rancio Sec made from Carignan, Macabeo, and all three colours of Grenache, aged for ten years. It is not muted, but with very ripe grapes has a natural 17.5% abv. Jean-François talked about *rancio* being part of the Catalan culture; families would keep wine in old barrels and his mother used it in the kitchen. He enthused about *gambas au rancio*. No 3 is a Vendange Tardive, from a field blend of both red and white grapes, aged outside in barrels of different sizes. It was elegantly honeyed with some raisiny notes. Next came a peachy Muscat de Rivesaltes followed by a 1998 Rivesaltes Hors d'Age, which had hints of prunes and nuts, and was rich, with wonderful intensity. La Grande Réserve came from a 25-year-old solera, which they began when they took over the estate. It was intense and rich, sweet but dry with great length. My tasting finished with a 1966 Rivesaltes Ambré, their oldest vintage, which was rounded and concentrated with a dry finish.

Vignerons Catalans

Perpignan

www.vigneronscatalans.com

Vignerons Catalans has long been one of the big players in Roussillon, but these days it is much less significant as the cooperatives have declined and the independent wine growers have soared in number. It was created in 1964, as a *groupement de producteurs*, to deal with the surplus of Vin Doux Naturel. Sales really developed in the 1970s, largely thanks to the French supermarket group Carrefour, which was started around the same time. The two companies grew together and in 1986 Vignerons Catalans included 51 of the 90 cooperatives in Roussillon, as well as 74 independent estates. By 2000 the number of cooperatives had shrunk to include 25 of the 28 cooperatives in Roussillon, with 2,500 *viticulteurs* farming 15,000 hectares of vineyards.

In late 2019, I visited Vignerons Catalans for the third time, in its extensive premises in the suburbs of Perpignan. The number of cooperatives has shrunk yet further, and now includes just eight, in the villages of Estagel, Baixas, Rivesaltes, Vinça, Rasiguères, Terrats, Trouillas, and most recently the Cave de l'Etoile in Banyuls, with a total of 1,500 wine growers. Together they cover 10,000 hectares, making Vignerons Catalans responsible for half the vin sec produced in the department.

These days Vignerons Catalans is classified as a Groupement d'Intérêt Agricole and its work is focused on conditioning, rather than blending, but with a team of competent oenologists. Above all they are a marketing and sales organization. They have various brand names and also work with French supermarkets, which remain important customers, for own labels. The export market is now much more significant, developing in Asia and Canada, and concentrating on Pays d'Oc, Côtes Catalanes and the appellations for vin sec, rather than Vin Doux Naturel. However, the cooperatives are increasingly independent. The cooperative at Terrassous, for example, sends a small part of its production *en vrac* for use in some supermarket lines, but otherwise it is quite autonomous. Vignerons Catalans also has partners in the Languedoc.

I have Cédric Haague, one of the winemakers in Perpignan, and responsible above all for bulk wine, to thank for a comparative tasting of the four villages of Côtes du Roussillon Villages, along with Les Aspres.

No one else produces all five Côtes du Roussillon Villages wines. The four villages are only ten kilometres apart, and the wines are all vinified in the same traditional way, but the various terroirs are quite different. In the early days of Vignerons Catalans, carbonic maceration was an important method of vinification, but Cédric observed that he has only used it once in 20 years. The blend in each of the wines is varying quantities of Syrah, Grenache and Carignan, depending on the harvest, and the levels of ripeness. The cuvée of Les Aspres comes from *galets roulés*, with a small amount of ageing in barrel, making for some rounded spice, with medium weight and a certain tannic streak. Lesquerde, grown on gneiss, had fresh red fruit with some elegant spice and supple tannins; it was my favourite of the five, and Cédric's too. Latour-de-France, from schist, had firmer spice with more weight and a firm tannic streak. Tautavel, from limestone, was quite solid, with firm, leathery fruit and more weight and structure, and Caramany from granite, with a fresher terroir, was rounded and perfumed with some ripe fruit and silky tannins. Cédric enthused about the spice of Roussillon, especially at Lesquerde. Over the last 20 years, there has been a drive for quality, with older vineyards and riper grapes, and a focus on traditional grapes. He loves the Agly Valley for its expression of Grenache and Carignan, particularly in the old vines that are the trump card of Roussillon. One does not encounter such old vines in the Languedoc.

10

THE TÊT VALLEY AND LES ASPRES

CORNEILLA-LA-RIVIÈRE

Domaine Boucabeille

Corneilla-la-Rivière

www.domaineboucabeille.com

Between the villages of Pézilla-la-Rivière and Corneilla-la-Rivière there is an avenue of trees that lean dramatically, amply demonstrating the power of the wind in Roussillon. Jean Boucabeille's cellar is close by. Altogether he has 40 hectares, all farmed organically, with vineyards on the lower slopes of Força Réal, so the terroir is grey schist. Geologically Força Réal is in layers, with schist on top, and vineyards at altitudes between 150 and 400 metres. There is talk of Força Réal as a village appellation, including the nearby villages of Corneilla-la-Rivière, Millas, Montner and Calce. But for the moment that is no more than an idea.

Jean began working with his father, whose family came from the Aude, near Narbonne. They were a family of *viticulteurs*, who sent grapes to the cooperative and were very much involved in the growers' unions. Jean's father, Régis, worked in Benelux, representing the Vignerons Catalans, Sieur d'Arques and the Cellier des Templiers and then in 1990 bought his own vineyards. He had no cellar and the wine was made at the cooperative in Corneilla-la-Rivière. Jean took over in 2001 and now has an attractive, modern cellar, with arches opening on to schist rock. He is bright and articulate, and talks perceptively about both his wines and Roussillon, with engaging enthusiasm. Régis, now in his 80s, helps with the harvest each year.

Régis B Les Terrasses Côtes du Roussillon Blanc is a blend of Grenache Blanc and Roussanne with floral fruit. Régis B Les Terrasses Rouge Côtes du Roussillon Villages, from Grenache with some Syrah aged in vat, has fresh, ripe fruit. Monte Nero Côtes du Roussillon Villages is made of equal parts Grenache and Mourvèdre aged in 300-litre barrels – you no longer need three varieties for Côtes du Roussillon Villages, as you once did. The oak was nicely integrated, with a tannic streak. A recent addition to Jean's range are some wines made without sulphur. The first, Amethyst – so called as the stone allegedly prevents drunkenness – is a pure Grenache Noir made with a classic vinification of destemmed grapes and aged in vat. The aim is to express the terroir as purely as possible and the wine had the spice of Grenache with some cinnamon notes and a firm finish. The second natural wine, Le Bon Sauvage – the name being a nod to Jean-Jacques Rousseau: *c'est sauvage sans sulphur* – comes from equal parts Carignan and Grenache, fermented together. This had rounded, ripe fruit and more weight than Amethyst.

I was treated to a vertical tasting of Jean's best cuvée, Les Orris Rouge. It includes at least 50 per cent Syrah each year with some Mourvèdre and some Grenache. The proportions depend on the vintage. The 2017, tasted in the autumn of 2019, was youthful and smoky, while 2016 was dense and more concentrated. The year 2015 was an excellent one, one of the best Roussillon vintages, with elegant balance, combining the fruit of 2017 and the tannins of 2016. The 2007 had notable Syrah spice, while, despite being quite a wet year, the 13-year-old wine from 2006 was drinking beautifully, balanced and supple. Finally, 2005 was elegant and balanced with a supple finish. The wines indisputably demonstrated the ageing potential of Roussillon.

Our tasting finished with a pair of white wines. Rêve d'Amour is pure Grenache Blanc, kept in foudres for three months, with some richness and a hint of vanilla. The wine had undergone a malolactic fermentation; Jean thought it would lose acidity as a consequence, but found it did not. Orris Blanc, from Grenache Blanc, Grenache Gris and Roussanne, with an élevage in barrel, was nicely balanced, with ripe fruit.

Jean's range of Vins Doux Naturels comprises a Rivesaltes Ambré and a Rivesaltes Tuilé, which he observed is generally made less frequently than Ambré. His is a *tuilé de saignée, muté sur jus,* with less than 100 grams per litre of sugar, with the aim of achieving a more modern flavour. An Hors d'Age, kept in old barrels for an average of 20 years, had delicious walnut fruit, and a new addition to the range, a Vin Sous Voile, Vin de France, was an intriguing combination of flavours, with notes of orange and honey and a firm streak of nuttiness. It is a pure Macabeo, aged for five years in barrel, without any *ouillage*, so that a *voile* develops, but it is less austere than a Rancio Sec.

Domaine Lafforgue

Corneilla-la-Rivière

www.chateau-lafforgue.fr

Noël Lafforgue had a grandmother from Calce and a grandfather from Corneilla-la-Rivière. His parents inherited the château in Calce and he was brought up in the village, while his uncle ran the family vineyards. Noël originally set out to be a musician, but changed his mind, studying at Carcassonne and learning from his uncle. 'I immersed myself in wine,' he recalls, and he has seen the changes in Calce. Noël now has an estate of 20 hectares, 15 in Corneilla-la-Rivière and 5 in Calce. However, he decided that it would be simpler to make his wine in Corneilla-la-Rivière, where he has a spacious cellar on the main street. His courtyard is shaded by an old plane tree that was planted when the cellar was built in 1911, and there are barrels outside for Vin Doux Naturel.

His uncle made virtually nothing but Vin Doux Naturel, but Noël realized that he needed to change. His first vintage was 2001, when he produced 400 hectolitres of Vin Doux Naturel and 100 hectolitres of vin sec. These days Vin Doux Naturel accounts for 10 per cent of production, with a little Muscat de Noël and occasionally some oxidative Vin Doux Naturel.

There is an easy-to-drink range of IGPs, a fresh, grapey Muscat and a rosé from all three colours of Grenache, *complanté* by his grandfather between 60 and 80 years ago. Côtes du Roussillon Les Feches, meaning a drystone terrace wall, comes from Grenache Blanc, with just 15 per cent Macabeo, fermented in barrels, with oak and fruit on the palate. Château Lafforgue Côtes du Roussillon Villages Les Courbatières, meaning *corbeau* or crow, the name of the *lieu-dit*, is a blend of Carignan, Syrah and Grenache, aged in 400-litre barrels, with ripe fruit and a streak of tannin making for an elegant balance. Noël is looking for freshness and elegance, recognizing that his wines have become more refined with experience. In 2017 Noël produced his first vintage of Plane d'en Bourgat, a Côtes du Roussillon Villages from 60 per cent Syrah and 40 per cent Carignan, aged for 12 months in barrel. He intends to make it only in the best vintages and was undecided about 2019.

Coming from very ripe grapes, it was rich and rounded, with vanilla from the oak; for me it was *un peu* too much.

Muscat de Noël, which is released on the third Thursday of November, was rich and concentrated with perfumed fruit and a hint of aniseed. A 2004 Ambré, bottled in January 2018, had spent four years outside in a barrel, as well as time in a concrete vat. It was a blend of equal parts Grenache (mainly Grenache Blanc) and Macabeo. With a sweet nose, but a firm palate and underlying sucrosity and hints of walnuts and orange, it was a delicious note on which to finish a week in Roussillon.

MILLAS

Domaine Força Réal

Millas

www.forcareal.com

The cellars of Domaine Força Réal, situated high on the slope of Força Réal, are worth the trip for their stunning position and breathtaking views. The track climbs steadily from a discreet turning near the village of Millas. I first visited this estate for an earlier book at the end of the last century, when I met the estate's creator Jean-Paul Henriques, who in 1989 opted to make his own wine rather than continue his father's négociant business. These days Jean-Paul is retired so I met his vivacious daughter, Laura, who has recently returned from working in Paris. Her brother Cyril works with her.

The building dates back to the mid-nineteenth century. It was restored by Jean-Paul and looks a little like a Mexican hacienda. A barrel cellar was added in 2007 and we tasted in what were once the old stables. The property includes 30 hectares of vines, farmed organically since 2014, all in one large holding on the slopes of Força Réal, at altitudes of between 100 and 450 metres. They also have 10 hectares of olive trees and want to develop their production of olive oil. There is a mill at nearby Corneilla-la-Rivière.

Le Blanc, Côtes du Roussillon, a blend of 80 per cent Macabeo with some Grenache Blanc and Viognier, has a peachy nose and some fresh, pithy fruit. Mas de la Garrigue, mainly from old Carignan with some Syrah and Grenache, is aged in stainless steel for eight months and has some spice and a fresh finish.

Domaine Força Réal Schistes is based on Grenache with some Syrah and Mourvèdre, and sees eight months of élevage in stainless steel. It has some fresh red fruit. The Grenache makes it more silky than the previous wine. Les Hauts de Força Réal Schistes is just that, coming from the vines at a higher altitude, namely the best Syrah, at 250–300 metres, with a little Grenache and Mourvèdre. With an élevage of 12 to 24 months, depending on the vintage, the wine was rich and rounded, nicely harmonious with dry spice. They still produce a little Rivesaltes Hors d'Age, with small barrels alongside the larger barriques in the ageing cellar, but it is no longer as important as it once was.

LE SOLER

Château Nadal Hainaut

Le Soler

www.chateaunadalhainaut.fr

You approach Château Nadal Hainaut down a magnificent avenue of plane trees, which were planted during the time of Napoleon III, like so many of the avenues of plane trees in France. At the end is a house built in the 1850s, with an extensive visitor area for tasting, as well as a Cistercian chapel that has recently been restored to charming simplicity. The attractive old cellar was built with the traditional narrow red bricks called *cayroux*, which are usually 44 x 22 centimetres but just 8 centimetres thick.

Jean-Marie Nadal is the sixth generation on the property, and his daughters work with him. The family arrived here in 1826 and in 1900 Jean-Marie's grandfather, François Nadal, married Thérèse Hainaut, hence the double-barrelled name. Altogether they have 45 hectares of vines, most of which have been farmed organically since 2010. The remaining 5 hectares constitute a vineyard of Carignan that was planted in 1900. It is very difficult to weed this vineyard mechanically as the vines are an unruly sprawl, with many so low that branches are resting on the ground. A tractor would be completely impossible, but there is no way that they would consider replanting the vineyard. The solution came in the form of a separate company hired to run that particular vineyard, since you cannot have organic and non-organic grapes vinified in the same cellar.

They make an extensive range of wines, including Muscat Sec, Chardonnay, both oaked and unoaked, and a pair of rosés, again oaked and unoaked. Possibly their most interesting red wine is Le Centenaire, from the venerable Carignan planted in 1900. There is some initial oak, with some sturdy fruit and rustic spice and the appealing fresh finish of Carignan. Jean-Marie observed that in the 1970s the estate consisted of 80 per cent Carignan, which was vinified by carbonic maceration. Then came the move towards Syrah, and also Cabernet Sauvignon, prompted by the demands of a more international market, so he pulled up most of his Carignan. Fortunately, he kept the now centenarian vines. Other reds include Terre de Quarante, mainly from Syrah with some Grenache, aged in oak. Signum is a selection of the best Syrah, with some dense fruit. Our tasting finished with a delicious Rivesaltes Ambré, a blend of Muscat and Macabeo aged in barrel and *bonbonnes* for 12 years.

LES ASPRES

The most recent addition to Côtes du Roussillon Villages covers vineyards in 19 villages between Elne and Castelnou, with about 35 producers. It is a hilly region, with small valleys, and modest altitude. The area has been very carefully delimited, so not all the vineyards in a village are included and there are numerous small plots surrounded by *garrigue*. Ideally, growers would like the specific village mentioned on the label as well, such as Les Aspres Tresserre, but that could be problematic if the vines are not actually in Tresserre, even though the cellar is. It would depend on the place of production. The appellation was created in 2017, with a retrospective use for 2015 and 2016, and requires three grape varieties in the blend, which differentiates it from other Côtes du Roussillon Villages.

CORNEILLA-DEL-VERCOL

Château de Corneilla

Corneilla-del-Vercol

www.jonqueresdoriola.fr

William Jonquères d'Oriola is the twenty-seventh generation of his family in the region. He knows his family has been at Corneilla-del-Vercol since 1337 and has been making wine since 1485, a date that is resonant with English historians for the Battle of Bosworth Field and the beginning of the Tudor dynasty. William is very humble; he sees himself as a link in the generational chain, and aspires to be a strong one. He has been working with his father, Philippe, since 2010, and in 2020 took over full responsibility for running the estate. As for the next generation, William's two small sons are already showing an active interest in tractors.

Back in 1987 when I first met Philippe he was one of about 20 independent wine growers in Roussillon, when production was dominated by village cooperatives and the large *groupements de producteurs* that had replaced the traditional négociants. Back then Philippe had a more international view of things than his colleagues: he studied oenology in Burgundy, before working in South Africa and then the London wine trade, returning to run the family property in 1970. William too has an international perspective, studying winemaking in Montpellier and going to business school in Nancy, before spending three years as a flying winemaker in Tasmania, the Margaret River and Marlborough.

The imposing red stone château was built by the Knights Templar, and the Jonquères d'Oriola family, who are Catalan in origin, settled there after the destruction of the order. William and I met in what was once the old cowshed and is now a spacious tasting centre for visitors. The biggest change since my last visit is the development of the vineyards, for the estate has grown from 58 hectares in 2010 to 95 hectares today, with a programme of replanting. The vineyards are in four main areas with different terroirs: near the coast between Cabestany and Canet; in Les Aspres between Bages and St Jean-Lasseille; in Collioure, as well as vineyards immediately around the village. In addition, they farm 6,000 olive trees. They follow the principles of HVE, which William considers to be even better than organic farming.

Our tasting began with Côtes du Roussillon Gris-Gris Le Rosé Porte-Bonheur, or lucky charm. It is a cheerful rosé, a blend of Grenache Gris and Grenache Noir with a little Syrah, picked in the early hours of the morning, starting at 3 a.m. and finishing no later than 10 a.m. so that the grapes are cool when they arrive in the cellar. Next came Heritage, another rosé, from Syrah with some Grenache Noir. The pale-coloured wine had some raspberry fruit on the nose and palate, with some weight. William favours a malolactic fermentation, which adds this weight, and retains acidity with an early picking date. Cavalcade Rosé followed, a blend of Grenache Noir and Mourvèdre, for which the malolactic fermentation had taken place in an American oak barrel, adding a hint of vanilla and some rounded mouthfeel.

Heritage Blanc is a blend of Macabeo, Grenache Blanc and Vermentino, with an élevage on the lees which made for some fresh herbal fruit and a rounded finish. The white Cavalcade, from the same varieties, spends time in oak, which gives more depth and nuance. The oak was well-integrated. Col de Mollo 231m is their Collioure, from vineyards at that altitude; the wind here means there is very little need for any treatments. A blend of Grenache Gris, Grenache Blanc and Roussanne, it sees what they call a Burgundian style of vinification, with a fermentation in 500-litre barrels. The Roussanne adds acidity and the altitude gives freshness.

Next came red wines. The Côtes du Roussillon Heritage is from Syrah with some Grenache and Carignan, fermented in vat. They keep the lees in suspension which makes for some rounded, supple fruit and spice. Château Corneilla Pur Sang comes from Syrah, Grenache, Carignan and Mourvèdre. With some ageing in barrel, it has ripe fruit and notes of tapenade. The name Pur Sang, or thoroughbred, is a reference to the family's other interest, breeding horses for show-jumping. William's great-grandfather was a world show-jumping champion in the 1930s and his great-uncle was twice Olympic champion.

Cavalcade Côtes du Roussillon Villages Les Aspres is a blend of 50 per cent Syrah with equal parts Mourvèdre and Grenache making up the rest of the blend. Elevage is in demi-muids. It was elegant and structured with satisfying depth. The Collioure is a blend, unusually, of 60 per cent Mourvèdre, with some Grenache and Syrah, aged in demi-muids for 12 months. Mourvèdre ages slower than Grenache, and this was fresh and tannic with some red fruit. The final wine was Côtes du Roussillon L'Indigène Sans Sulfites Ajoutés, a blend of 80 per cent Grenache with some Mourvèdre, from a 2.7-hectare plot near Bages in Les Aspres. When they bought the plot in 2018 its 96-year-old former owner had barely touched it for ten years. The vines were a jungle but they had not endured any chemical treatments. William explained that he was looking for a different profile, 'such as my great-grandfather would have made', and the flavours were ripe and fresh, with an elegant finish, the result of a carbonic maceration.

As for Vin Doux Naturel, as elsewhere in Roussillon, its importance has greatly declined. At his first vintage in 1970 vins secs already accounted for 75 per cent of Philippe's production, unusual for the time; today almost all the wine is vin sec and Vin Doux Naturel makes up just 5 per cent of production. William, gently encouraged by his father, exudes a sympathetic energy, which promises well for the future of the estate.

PASSA

Domaine Puig-Parahy

Passa

www.puigparahyvineyard.wordpress.com

Passa is in the Les Aspres but my visit to Domaine Puig-Parahy concentrated on Vin Doux Naturel. Georges Puig comes from an old Roussillon family and so we talked history, about which he is passionate, both his own family history and the history of Vin Doux Naturel: 'The transmission of history must not stop,' he says. An early family will obliged one son to provide his mother with a quantity of wine, while his brother inherited the silver drinking service. Not so long ago his grandfather was the second largest producer of Vin Doux Naturel, after M. Bourdouill at Château de Jau. Georges learnt his winemaking from his father and grandfather, observing that his grandfather made wine *à l'ancienne*. We sat in the rather formal drawing room of the family home, just opposite the church in Passa, before visiting the cellar, which dates from the fourteenth century.

The cellar was a veritable Aladdin's cave, with the intoxicatingly pervading aroma of maturing Vin Doux Naturel. There was a large wooden trough, used for crushing the grapes, the remains of a dismantled eighteenth-century press, and numerous old barrels of varying sizes. Georges still has an old label for a wine made by his cousin, Dr François Puig, at Domaine de San Lluc in Passa, labelled 1936 Côtes du Haut Roussillon. Back in the drawing room we tasted some truly venerable bottles, beginning with a 1989 Rivesaltes Tuilé, based on Grenache Noir with some Carignan, which was rich and nutty. A 1982 was only recently bottled, and was quite rich and sweet, with a firm bite. It was muted at 100 grams per litre residual sugar. Georges observed that in the nineteenth century the wines were much sweeter, closer to 150 grams per litre residual sugar, depending on family taste and tradition. He prefers 93 grams per litre, and these days is making more Rancio Sec. A 1970 Rivesaltes Tuilé, again recently bottled, was rich, with intense concentration. Next came a 1968 Ambré, from Grenache Blanc and Macabeo. Georges explained that you never clean the barrels, always retaining the mother, rather like a Vin Santo, and if the barrel is defective, you transfer the mother to another barrel. Our tasting finished with a 1945 Tuilé, which was dry and intense. The sugar had long since disappeared, leaving an incisive, razor-sharp, salty acidity and fruit. I quizzed Georges about his winemaking methods: laissez-faire would describe them most accurately. It was an enormous privilege to taste these wines, and I felt that I had stumbled into a time warp. Sadly, Georges has no children to take over the estate and he described the viticultural situation of Roussillon as disastrous, particularly compared to other places that produce fortified wine such as Spain, Portugal or Samos. A wine estate was once worth a fortune, but this no longer holds true.

PONTEILLA

Mas Bécha

Ponteilla

www.masbecha.com

Charles Perez took me for a drive through his vineyards, with their dramatic backdrop of the Pyrenees. He owns a large expanse of land, some 110 hectares, in the hamlet of Nyls, of which 30 hectares are vineyards. He also has olive and almond trees, cork oaks and fields for cereal. Charles' father bought the estate in 1997 and Charles has been involved since 2003, after studies in Carcassonne and practical experience in Bordeaux and Texas. Although his father worked in transport it was, as Charles put it, fashionable to own a wine estate. The wine was sold *en vrac* until Charles took the decision to bottle his wine. He is successful on the Chinese market, so that he is 'no longer a slave of the *négoce*'.

Ponteilla, of which Nyls is a part, is the last village of Les Aspres before Perpignan. Essentially, Charles has two main soil types, one based on sand, and the other on clay. Grenache is good on sand as it ripens without too much alcohol, while clay is a hotter soil, making for more alcohol. He also has Syrah and Mourvèdre, but no Carignan. The vineyards are on a series of terraces, and are mainly bush vines so that the grapes are protected from sunburn. Charles enthused about making his own compost and is experimenting with different mixes of ground cover.

Back in the cellar, each plot is vinified separately; Charles looks for what he called *îlots* or islands of homogeneity. He likes a cool pre-fermentation, especially for Syrah, which gives density and weight without alcohol. He does some *délestage* and very little *remontage*, as he wants to avoid too much extraction. Essentially, he wants velvet tannins which emphasize the fruit. We wandered round the cellar, glass in hand, tasting from vats and barrels, with Charles expounding his ideas. He is very thoughtful and articulate with a vision of his objectives: you sense a pioneering spirit.

Charles makes a varied range of wines, both sec and doux, including some individual vineyard selections. The Côtes du Roussillon Excellence is based on Syrah, with fresh ripe fruit and a supple streak of tannin. The freshness comes from the natural acidity, with the work on the tannins enhancing the impression of freshness. In addition, he makes a Côtes du Roussillon Villages Les Aspres Classique from Syrah, Grenache and Mourvèdre. Les Aspres Barrique Serge, named after his father, completes the range. My tasting finished with La Bergerie de Camps de Nyls, a Côtes du Roussillon Villages Les Aspres, which is also based on Syrah. It was richer and more intense than the Excellence, with black fruit, but still had a refreshing finish. Charles is still searching for the style of his white wine, for which he has Vermentino, Grenache Blanc and Macabeo, which is picked early to retain the acidity. It is fermented in barrel, with regular *bâtonnage* and had hints of aniseed.

As I left I admired an old mulberry tree by the cellar door, which may be a cutting from a tree planted at Versailles at the time of the Revolution.

Château Lauriga

Traverse de Ponteilla, Thuir

www.paulmas.com and www.lauriga.com

Domaines Paul Mas, headed by the energetic Jean-Claude Mas, is better known for its Languedoc wines. However, Jean-Claude extended his activities into Roussillon with the purchase of Château Lauriga in 2015. It is an attractive Catalan mas, a large *batisse* built in the nineteenth century from the distinctive *cayrou*, the thin red bricks of Roussillon, with 56 hectares of vines on relatively flat land near the town of Thuir. The name comes from Laurinya, *laurier rose*, or oleander. The estate is gradually being converted to organic viticulture, with some vineyards being replanted.

David Costa, the winemaker when I visited in 2019, comes from Roussillon and studied in Montpellier. He worked for Domaines Paul Mas for some 15 years, originally in the Languedoc, and then at Château Lauriga, but has since moved on to pastures new. They have kept the original range of the previous owner, but with what David called 'la griffe de Jean Claude'. Lauriga Macabeu is firmly stony, while Soleil Blanc de Lauriga, with some Grenache Blanc as well as Macabeo, has some obvious oak. David observed that Macabeo is currently quite fashionable; the flavours are neutral, but it adds vivacity. Muscat is emblematic of Roussillon, but that is a handicap. There is too much of it, 'il faut le dépoussiérer', said David – dust it down. A pair of rosés were quite deep in colour: 'you cannot have pale rosé in Roussillon', points out David. As for the reds, Domaine Lauriga Racine is a pure Carignan, including some very old vines. It was nicely structured with ripe berry fruit. A trio of Côtes du Roussillon had variations in their blends and élevage. Laurinya was the most structured, and a new addition to the range in 2016, with a bottle that is a copy of a crystal bottle from a Parisian glass producer. David commented that you do not need wood if you have ripe grapes; the fruit is sufficient.

To his taste the region's Vin Doux Naturel has too much sugar. His own Rivesaltes Grenat was fresh and rounded, the Muscat de Rivesaltes was fresh and slightly honeyed and the Rivesaltes Ambré Hors d'Age was firm and nutty with just a hint of sweetness on the finish. David enthused about the region's potential but mourned what he saw as a closed-minded attitude and an unwillingness among many producers to change. Château Lauriga is in good hands under the new owners, with projects to develop wine tourism.

TERRATS

Domaine Ferrer-Ribière

Terrats

04 68 53 24 45

The cramped cellars of Domaine Ferrer-Ribière are in a narrow back street of Terrats. Bruno Ribière explained how it all started, when he met Denis Ferrer back in 1993. He was a retired office worker and he described Denis as a 'viticulteur débutant'. They began working together, and are now helped by Bruno's daughter, Mireille, who will eventually take over the estate. They have 18 hectares, in 33 small plots, all around the village of Terrats, all farmed organically, and all planted with the traditional varieties: Grenache, Carignan, Syrah and Mourvèdre for red wine, and for white wine Grenache Blanc, Grenache Gris and Macabeo. Bruno enthused about the terroir as well as the culture and the Mediterranean cuisine. The wines all start as single varieties from a specific vineyard, and may then be blended. As for Vin Doux Naturel, Bruno says they are fabulous, but too complicated commercially so they have stopped making them, though Bruno did admit to making some Muscat 'for our retirement'.

Bruno is very articulate and has a wonderfully expressive, bearded face. He was sporting a T-shirt bearing the apt slogan 'Vignerons du Roussillon' and gave me a very generous tasting, despite the imminent arrival of their pickers for their harvest lunch. Cuvée F is a blend of Grenache Blanc with 10 per cent Muscat, with rounded fruit and a herbal note. A plot of 92-year-old Grenache Blanc Côtes Catalanes was given an oxidative Catalan élevage in demi-muids so that it was light orange in colour. The juice was kept in wood for up to two years without any *ouillage*, and sulphur added only at bottling. It was very intriguing. A pure Grenache Gris Côtes Catalanes was also aged in barrel, with some rounded fruit and balancing acidity. Bruno enthused about *rancio*, describing it as 'la fibre catalane' and noting that it is fundamental to Roussillon society, traditionally drunk outside in the village on a summer's evening, but it is also a great food wine. Sans Interdit, Macabeo Côtes Catalanes from a solera begun in 2001 was amber in colour, firm, austere and mineral – 'c'est violent; c'est vrai,' says Bruno – and just the thing to go with anchovies or manchego cheese.

Bruno described Carignan as 'le cépage roi pour les vins sec'. Empreinte du Temps comes from 142-year-old Carignan vines. These are mentioned in his great-great-grandfather's cellar book in 1878 because he had paid a workman to graft them onto *labrusca* vines prior to the arrival of phylloxera in Roussillon. The wine was redolent of fresh red fruit, with spice and balance. Bruno observed that he did everything as simply as possible. A Grenache Noir was rounded and spicy, with a touch of cinnamon; a Mourvèdre was quite sturdy. A fresh, peppery Syrah came from an egg made from a particular material that breathes like wood, but without any aromatization or effect on flavour. A second Carignan, largely from 120-year-old vines, was quite sturdy and quite different from the first wine. Bruno enthused about Carignan, considering that it would be much better to concentrate on Carignan, as well as Grenache, since Syrah, he believes, is not suitable for Roussillon. On the other hand, Mourvèdre, or Mataro, is Catalan. Cuvée F Côtes Catalanes, from Carignan and Grenache, had lovely fresh fruit, while a Côtes du Roussillon Tradition, from all four red varieties, was sturdier, with ripe fruit. Côtes du Roussillon Villages Les Aspres Caṇa, based on Mourvèdre with an élevage in eggs, was rounded and elegant with a fresh finish.

Next came a pair of Muscat de Rivesaltes. The younger 2017 was fresh and rounded, while the 2014 had some *rancio* notes with a hint of oak after four years of ageing. My tasting finished with a Rancio Vin de France Perle d'Octobre, from a Grenache Gris and Grenache Blanc solera started in 1994. The grapes were dried for a month on straw, and the wine was long and nutty with great concentration, and indeed not unlike a straw wine.

Vignobles de Constance et du Terrassous

Terrats

www.terrassous.com

This is a group of three village cooperatives, with cellars in Terrats, Thuir and Fourques, and vineyards mainly in Les Aspres, in the villages of Fourques, Montaurol and Ste Colombe-de-la-Commanderie. A Terrassous is an inhabitant of Terrats, which means, literally, land that produces money (*la terre qui produit de l'argent*). Altogether they have 80 members, with 700 hectares. I met the previous director, Hervé Lasserre, in Fourques where they have a smart tasting room, but Hervé particularly wanted to show me their barrels, in an extensive area filling the whole attic of the large cellar. It is not air-conditioned so the barrels, as many as 772, are subject to all extremes of temperature. A momentary power cut enhanced the atmosphere, as we wandered among the barrels in the gloom. They still do a traditional *mutage*, treating every barrel individually. The barrels are never topped up, so that 3–5 per cent evaporates each year, meaning that the contents of a small barrel would eventually disappear entirely over 45 years. Their oldest vintage is 1974. Hervé explained that the evolution is not linear and they taste regularly to decide what to bottle. The wines are delicious, and sell much better abroad than in France itself, mainly to the US and Asia. A six-year-old Ambré, from Grenache Blanc, was rounded and lightly nutty; the 12-year-old version was dry with firmer fruit and at 18 years old the wine had become deep in colour with incisive fruit and acidity. The 1981 was austerely nutty, and my tasting of the Vin Doux Naturel finished with the 1974, with more weight and notes of prunes. As well as Rivesaltes, there was Muscat de Rivesaltes, for which they prefer Muscat à Petits Grains, muting at 110 grams per litre residual sugar to obtain 16.5% abv, aiming for youth and elegance. Despite their extensive barrel cellar for Vin Doux Naturel, it still only accounts for a small part of their production; 50,000 bottles out of a total annual production of 350,000.

Hervé enthused about Les Aspres for its freshness. Most of their wines could fit into the appellation, but they produce an wide range of

Côtes Catalanes as well. Their technical director Emma Nieto Charquès is working on site selection. They have a lot of old Grenache, but it is less expressive in Les Aspres than in the Agly Valley, so they tend to use it for rosé as it lacks the concentration for red wine. They have some good vineyards of old Carignan. Highlights among their red vins secs included a pure Carignan and various Côtes du Roussillon and Côtes du Roussillon Villages Les Aspres. A Côtes du Roussillon Réserve was ripe and spicy, intended for easy drinking. Château Mossé Côtes du Roussillon, with an élevage only in tank, was ripe and rounded, with black fruit. Les Pierres Plates Les Aspres had ripe spice balanced with a firm streak of tannin and SuMMum was a blend of the best sites with lower yields. The oak was well-integrated.

I left with the impression of a well-run business that was focused on future quality, while retaining contact with the traditional.

TRESSERRE

Clos Mané

Tresserre

06 66 64 98 34

Following a career with Airbus in Toulouse, Baptiste Milhes Poutingon took over the family vineyards outside Tresserre, with a first solo vintage in 2017. The family came originally from the Vaucluse but left for Algeria when their vineyards were destroyed by phylloxera in the late nineteenth century. When Baptiste's great-grandfather returned to France he bought the farm at Els Barbats. He clearly had no intention of moving again – Clos Mané means *je reste*, in other words, I have put down my roots. Originally the estate produced Vin Doux Naturel, which was sold to the *négoce* but then Baptiste's father, Paul, replanted the vineyards to concentrate on vin sec, and began to bottle some in 1995. He retired in 2016 but still retains a keen interest, while giving his son a free rein. Tasting with father and son, you enjoy a lively repartee, with good humour shared between the two generations. The estate now comprises 10 hectares of vines, all within Les Aspres, as well as nut and fruit trees, including olives, cherries, apricots and almonds.

Nowadays the focus of the estate is very much on vin sec, but there were some *bonbonnes* on the roof as well as others in the cellar, with Baptiste wanting to see what works best. A Côtes Catalanes Rosé Rosaline, named after a grandmother, was mainly Syrah, with some underripe Carignan to add acidity. It was fresh and lively. Their white wine, named after another grandmother, Blanche, is a Muscat, but a subtle interpretation of the variety. It is relatively low in alcohol, and with a long, slow fermentation and work on the lees, it has other more intriguing flavours. Marion, named after Baptiste's sister, is a Côtes du Roussillon blend of Syrah, Grenache and Carignan, all picked together and aged in vat, with supple fruit and freshness. Côtes Catalanes L'Impertinent, with more Syrah, also kept in vat, is denser and more peppery, with good ageing potential. Côtes du Roussillon Villages Les Aspres Les Vignes de Mon Père, is a blend predominantly of Syrah, with some Grenache and a little Carignan, the result of site selection, with Syrah grown on a north-facing slope providing freshness. It is aged in vat for two years, making for rounded, harmonious tannins and balanced fruit. Our tasting of vin sec finished with Carignan. First came a pure Carignan, from a plot near the house where Baptiste remembers playing as a small boy. It was a delicious interpretation of that often-maligned grape variety, with freshness and spicy fruit. Next, Paul generously opened his Carignan from his final vintage, 2016. It was denser and sturdier than Baptiste's, making a fascinating comparison of the two generations. Then came an oaked Carignan, Autre Chose, with a balance of tannin, freshness and well-integrated oak.

Baptiste is keen to explore new avenues and possibilities. He would like to plant Cinsault for both red and rosé, as well as Grenache Gris, Grenache Blanc and Carignan Blanc. Our tasting finished with Vin Doux Naturel, with a Muscat de Rivesaltes, Satin, which followed 'an old recipe from my great-grandfather', with equal amounts of Muscat d'Alexandrie and Muscat à Petits Grains. The Muscat à Petits Grains is picked when the grapes are really ripe, and the Alexandrie at a lower degree, and the *mutage* represents 7 per cent, not 10 per cent of the volume, so that the wine is both lower in alcohol and sugar than a more traditional Rivesaltes. Baptiste observed that people want less sugar these days and he would also like to try making a *rancio*. Rivesaltes Grenat Garance came next, with spice and ripe fruit, and finally a 2002 Rivesaltes Ambré that was all the better for having been forgotten in the cellar, with rounded nutty fruit. It made a delicious finale.

Domaine Vaquer

Tresserre

www.domaine-vaquer.com

Domaine Vaquer is one of the historic estates of Roussillon. I first met

Fernand Vaquer in the late-1980s when he made some of the most expensive wines of the whole of the south of France, bottling them under the humble category of Vin de Table, not even Vin de France. They were quite unlike anything else from Roussillon, with an intriguing originality that defied definition. Fernand's son, Bernard, and his Burgundian wife, Frédérique, took over the estate in 1989 but Bernard sadly died very prematurely in 2001. Frédérique has run the estate single-handedly ever since. Fernand, however, retained a lively interest in the estate until his death at the ripe old age of 89 in 2018.

Quizzed about the typicity of Les Aspres, Frédérique said balance, in both the red and white wines, although only the red wines qualify for the appellation. The soil is clay with *galets*. She has 18 hectares, including some Grenache planted in the 1950s and some Carignan from 1937. The battle of Le Boulou was fought nearby during the Napoleonic wars and she has found cannon balls in the vineyards. On a previous visit, I went for a walk in the vineyards. There should have been a great view, with the sea in the distance, but the Canigou was in the cloud. Frédérique referred to 'le toit de mon monde', the roof of my world, encompassing the village, the wine, birdsong, the Pyrenees and the Canigou, and views on all sides towards the hills of the Albères, the sea, Perpignan, Rivesaltes and the peak of Força Réal. She has always worked as organically as possible and began the official conversion in 2019. In practice, little will change.

I always enjoy tasting with Frédérique. She is thoughtful and sensitive, and fairly laissez-faire in her winemaking, using a little sulphur at the harvest, and prefers to 'let the wine get on with it'. In practice this means monitoring and adjusting if necessary; she sees 'if it is as it should be, going in the right direction or in need of help. You will have time to react. It's a bit like children; you give them an education and then let them choose.' There is an elegant Burgundian touch to her wines, which have texture and a satisfying mouthfeel. White Esquisse, based on Roussanne with some Macabeo, is fresh and floral. L'Exception Blanc, from Grenache Gris and Grenache Blanc, with some Macabeo, is aged in demi-muids for 12 months and has some rounded, resinous flavours and balancing acidity. To show just how well her wines age, she opened a 1991 Macabeo, which was rich and resinous, with dry honey and a salty finish. It was very intriguing with nuances galore.

Cuvée Bernard is a Côtes du Roussillon, a co-fermented blend of Carignan, Grenache and Syrah. Intended to be a *vin de plaisir*, it is wonderfully drinkable, with the elegance of Les Aspres. Frédérique observed that while the wines of Roussillon can be criticized for being too heavy she is adamant that they can be fresh. Next came L'Exigence, a blend of Grenache Noir and Carignan with ripe cherry fruit and then a pure Carignan from vines that were planted in 1991. Les Aspres L'Exception is a blend of Grenache, Carignan and a little Syrah that spent 18 months in a cement vat. For Frédérique this is the expression of Les Aspres. It is rounded with more depth, but not heavy. She does not want her wines too marked by élevage.

A couple of older vintages followed, first the 2011, which was ageing beautifully and then the 1988, Fernand's last vintage, a blend of 80 per cent Carignan with 20 per cent Grenache. Fernand used to age his wines at 1,300 metres at the foot of the peak of the Puigmal, which offered a different atmospheric pressure than at a lower altitude. It would have spent two years in vat and then 23 years in bottle, up in the mountains. It was the colour of Burgundy, with elegant, ethereal fruit and was still remarkably fresh and very much alive. Frédérique observed that Fernand only bottled a wine when he liked it. She finds it difficult to persuade people to try a 30-year-old wine from Roussillon, whereas they would not hesitate over a 30-year-old claret or Burgundy. There is no notion of vintage in Roussillon. Les Aspres Epsilon is mainly Syrah aged in barrel for a year. The oak reveals the flavours of the Syrah rather than masking them, and the wine was nicely concentrated, with peppery notes.

Then we moved on to Vin Doux Naturel, beginning with a Rivesaltes Grenat, L'Extrait, which had fresh ripe fruit. Next came Heritage 1986, a Rivesaltes Ambré Hors d'Age made by Fernand, who had inherited a solera of old wine. It was wonderfully nutty, Frédérique suggested *pain d'épice*, with balancing acidity and richness. The delicious 1994 Préface Rivesaltes Ambré Hors d'Age was drier with notes of orange, and was a splendid way to conclude the tasting.

TROUILLAS

Domaine de Treloar

Trouillas

www.domainetreloar.com

Englishman Jonathan Hesford and his New Zealand wife, Rachel Treloar, were living in New York on 11 September 2001 and saw the Twin Towers collapse. On that day their lives changed. Jonathan decided that the world of international IT was no longer for him, and that this was the moment for a complete change of career. They returned to England: a brief course at Plumpton College followed, and then an intensive post-graduate course of viticulture and oenology at Lincoln on New Zealand's South Island. Jonathan spent two years as assistant winemaker with Tim Finn at Neudorf in Nelson before the couple decided to return to Europe. Looking for a vineyard in the south of France they came to Roussillon, where, as Jonathan noted, they are both foreigners. They love the landscape, the beaches, the proximity to Spain.

In 2005 they found an enormous cellar in the village of Trouillas and bought 10 hectares of vines. Jonathan has subsequently modernized the cellar. There is a state-of-the-art pneumatic press, the concrete vats have gone, replaced with stainless steel tanks, and just one of the giant foudres remains. Domaine de Treloar is not certified organic, but Jonathan uses no weedkiller or fertilizer and sprays as little as possible, following the precepts of HVE. As an outsider, Jonathan is not afraid to voice his opinion on Roussillon. He is scathing about the creation of Les Aspres as a Côtes du Roussillon Villages. Côtes du Roussillon is made according to the INAO rule book, but has a lower bulk price than any appellation in the Languedoc, he points out. He is certainly aware of the cumbersome bureaucracy that puts a break on the development of the region's wines. In his view the IGP Côtes Catalanes has a nice cachet, sounding more exclusive, especially for white wines.

One Block Muscat, so named as their apartment was just one block away from the World Trade Center, is a pure Muscat à Petits Grains, with some fresh grapeyness. Daltonien is their newest wine, with some Muscat d'Alexandrie given some skin contact and some Muscat à Petits Grains without skin contact blended and kept in wood with some *bâtonnage*. The Alexandrie softens the flavours with more texture and nuances. La Terre Promise, mainly from Grenache Gris, with some Carignan and Macabeo, is fermented in barrel and has some buttery weight. Planète Rose is quite a deep-coloured pink, a food rosé combining fresh acidity and depth.

Le Ciel Vide, named after Bruce Springsteen's song, 'Empty Sky', composed after 9/11, is a blend of the 'leftovers', which in the 2018 vintage meant Mourvèdre and Grenache, for a wine with some spice. One Block Grenache is a blend of Lledoner Pelut and Grenache Noir, from two adjoining vineyards, with the refreshing perfume of unoaked Grenache. Côtes du Roussillon Three Peaks is a blend of Syrah,

Grenache and Mourvèdre, aged in tank and old wood, with some ripe spice and firm tannins on the finish. Côtes Catalanes Le Rescapé is a pure Carignan from 80-year-old vines, part of which undergoes carbonic maceration and part a traditional fermentation. Aged for nine months in old wood, the wine has firm cherry fruit, making it more suitable for summer drinking. Jonathan observed that Roussillon makes impressive winter reds but most tourists visit the region in the height of summer, and consequently completely ignore the region's red wines. He also wants to make red wines that will age. A pure Syrah, 2013 Le Secret was evolving nicely in the summer of 2020, while 2017 Motus, based on Mourvèdre, was firm and structured. The name comes from the expression *motus et bouche cousue*, or Mum's the word, that the wine is made from just one variety. Our tasting finished with Côtes du Roussillon Tahi, 2013, the name of which means 'one' in Maori. A blend of Syrah with some Mourvèdre and Grenache from selected vineyards, it was firm and sturdy with black fruit and plenty of ageing potential.

MONTESCOT

Château de l'Ou

Montescot

www.chateau-de-lou.com

Philippe Bourrier and his wife, Séverine, bought the Château de l'Ou in 1999. It was an old estate that had sent its grapes to the local cooperative for wine *en vrac*, so apart from the vineyards themselves there was nothing there when they arrived. Ou means 'egg' in Catalan, and is the name of the *lieu-dit*, so-called because of a spring on the land in the shape of an egg. Philippe has an agricultural background and Séverine is the winemaker, for about 70 hectares of vineyards. They have 25 hectares at the property outside the village Montescot, which is not one of the villages of Les Aspres, and the rest are in the Agly Valley, with 15 hectares in St Paul-de-Fenouillet, 27 in Maury and 10 in Caudiès, one of the coolest areas of Roussillon. They bought some Chardonnay vineyards from Jean-Marie Allery, who had established a reputation for this grape variety at the end of the last century.

With so many vineyards, they have an extensive range of wines, and I was treated to an appropriately diverse tasting, during which Philippe talked about their wines with thoughtful perception. Grenache Gris is vinified in amphorae; the grapes are destemmed and the juice 'stirs itself', 'se bâtonne toute seule – la jarre du fainéant,' quipped Philippe. He enthused about the purity of the grape variety with its minerality. L'Infiniment de l'Ou, a pure Chardonnay that has spent eight months in barrel with no malolactic fermentation, was lightly buttery and structured, with good acidity. Château de l'Ou Blanc was a fragrant blend of Grenache Gris and Grenache Blanc, along with some Roussanne. Côtes du Roussillon Esprit Libre is a blend of Syrah and Grenache aged in vat. 'We look for fruit,' observed Philippe, and they have achieved some rounded spice. They favour a cool pre-fermentation at 4°C and the grapes are crushed, with selected yeast. They first made this wine without any sulphur in 2018. The same wine with a small amount of sulphur was also nicely rounded, and perhaps slightly more tannic.

Altogether they make five different wines from Syrah. Côtes Catalanes Cuvée l'Ove is made in a *jarre*, with a three-week maceration, after which it is racked and then returned to the *jarre*, which gives the wine freshness, black fruit and a touch of pepper. Infiniment, with a *vinification intégrale*, was rich and solid with hints of chocolate, and for my palate *un peu* too much, as the French so elegantly say. Secret de Schistes comes from vineyards in St Paul-de-Fenouillet, which is cooler than Montescot, with a bigger diurnal difference. Also with a *vinification intégrale*, it was rich and rounded, but with a streak of freshness from the schist. Velours Noir, from Syrah grown in the cooler climate of Caudiès, was closed and peppery.

Compartir, meaning to share, is a blend of Syrah from St Paul-de-Fenouillet with Grenache and Carignan from Maury, with some élevage in wood and some in *jarre*. The flavours were fresh, with the *griottes* (sour cherries) of the Grenache balanced with tannin. Côtes Catalanes Grenache Rhapsody, a pure Grenache, comes from old vines in Maury and was vinified and aged in amphora. It was fresh and elegant, with pure fruit. Ipso Facto from a single plot in St Paul-de-Fenouillet was deep in colour with spice, peppery fruit and supple tannins.

My tasting finished with a Rancio Sec of Grenache Gris that had seen five years ageing, producing firm saline fruit and what Philippe called a *côté amer*, a bitter edge. They made it for the first time in 2006. My visit ended with a brief look at the spacious cellar, with its *jarres*, and barrels with brightly painted metal bands. They also have an egg made from oak, one of only two in France, the other being at Château Pontet-Canet in the Médoc. The wine moves in exactly the same way as it does with a concrete egg.

11

BANYULS AND COLLIOURE

Coming from Spain, Banyuls is the first wine of France you encounter. After the unprepossessing border town of Portbou, you take the coastal road past Cerbère, to Banyuls, passing Cap Réderis where you can enjoy a magnificent panorama of the coastline on a clear day. The road hugs the coast, offering dramatic views at every stomach-churning hairpin bend. There are the remains of old watchtowers, part of the warning system of beacons to raise the alarm against invaders. The Tour Madeloc dominates the skyline above Collioure. Here the Albères, the foothills of the Pyrenees, cascade into the Mediterranean, with the vineyards on steep terraces. The landscape is dominated by the terraces, with kilometres of drystone walls, some say as many as 6,000 kilometres, which perform a vital function in retaining the soil and preventing erosion. No one quite knows when they were first constructed; some may go back to the Romans, others are more recent. The winds blow hard and the vines cling tenaciously to the hillsides.

Banyuls is a cheerful, seaside town. The sculptor, Aristide Maillol, was born here in 1861 and there are several of his elegant sculptures of women on the seafront, the Allées Maillol, as well as at the Musée Maillol, situated in the house where he died in 1944, a little way out of the town. You will also find Maillol's work amongst the various war memorials, at Elne, Céret and Port-Vendres, as well as in Banyuls. Curiously *maillol* in Catalan means a young vine. The seafront is lined with restaurants and cafés, and in the back streets there are numerous wine growers' shops and cellars, and most conspicuously the large premises of the Cave l'Etoile.

Driving on past Port-Vendres and Cap Béar you come to the charming fishing village of Collioure. We stayed in a tiny fisherman's cottage on a steep, cobbled street in the heart of the town. One night there was a violent storm and we awoke to find a river gushing down the cobbles. The seafront is dominated by the imposing Château Royal, the summer residence of the Kings of Majorca, before Collioure became part of the Kingdom of Aragon. French military engineer Vauban added further fortifications in the seventeenth century following the Treaty of the Pyrenees. There is also an imposing fortified church and a tiny chapel on a small promontory. Collioure has long retained an artistic tradition; Matisse and Derain worked here and there are still artists' studios in the old town. The walls of the restaurant Les Templiers are covered with paintings, as the Pous family, now the third generation, has continued the custom of accepting canvases as payment for restaurant bills. Collioure is also known for its anchovies, which are an essential part of Catalan cooking, often allied with red peppers to make a delicious *salade catalane*. The tradition of the preparation of anchovies in Collioure goes back to the days when there was no tax on salt and people were both fishermen and wine growers. A wander through the old town is rewarding for little streets, the occasional wine cellar, and numerous restaurants. We enjoyed meals at El Capillo, with Catalan flavours, and Le 5ème Péché, 'the fifth sin' (gluttony, of course), which has a Japanese chef. Chez Syl Vins is a cheerful wine bar with an eclectic range of local wines.

The wines of Banyuls and Collioure come from identical vineyards for the two appellations cover the four villages of Banyuls-sur-Mer, Collioure, Port-Vendres and Cerbère. The area is called La Côte Vermeille, which is also the name of the local IGP, and comes from the colour of the rocks, with their high red iron content.

Essentially the difference between the two wines is one of vinification: Banyuls is a fortified Vin Doux Naturel and Collioure is an unfortified table wine. The locals simply refer to one as vin doux and the other as vin sec. Banyuls was one of the first appellations of France, created in 1936, while Collioure came very much later, with a red wine in 1971, rosé in 1991 and Collioure Blanc later still, in 2003. Until the creation of the appellation, Collioure was often called Banyuls Sec, and in the nineteenth century the two names were almost interchangeable, for both Banyuls and Collioure were praised as smooth, rich wines.

The Knights Templar had an important presence in the area during the Middle Ages, adopting the winemaking methods of Arnaud de Villeneuve, and Banyuls was enjoyed at the Aragonese court. However, the area remained something of a vinous backwater. In the nineteenth century Jullien cited Banyuls, Collioure and Port-Vendres in the second category of his classification of red wines, while Guyot described Banyuls as an exceptional wine and very much sought after. It was made solely from Grenache Noir, pressed immediately and muted with spirit. Cavoleau enthused about the good wines made from Grenache Noir in Banyuls, Collioure and Port-Vendres, which at eight to ten years were 'velvety and rich, but delicate and very agreeable'. The best wines were very dark in colour and could be very sweet. Rendu said of the wines of Collioure that they had a beautiful colour and body and a lot of richness, that they held the balance between fortified and dry wines and that when they aged, they acquired finesse and a pronounced bouquet. They should not be bottled for ten years, by which time they should have taken on a *rancio* character, he advised.

Michel Berta of Domaine Berta-Maillol is well informed about the history of Banyuls. He related how insular Banyuls was in the middle of the nineteenth century. There were tracks for donkeys, but the easiest form of transport was by sea, with Port-Vendres the main port. It was a region of sailors, fishermen and vignerons. There were commercial ties with North Africa and viticultural links too, as for both regions Grenache Noir was the important grape variety. Michel's great-grandfather bottled the domaine's first wine, but it was easier to ship in barrels than bottle. Before the First World War most wine was sold to the *négoce*. Michel commented that the impact of the First World War forced a French identity on the region.

Vincent Cantié of Domaine La Tour Vieille also talked about his family. His grandfather had earned enough money to invest in vineyards; his father fished for anchovies in the summer and worked his vineyards in the winter. Vin Doux Naturel was profitable between the two world wars and immediately after the Second World War. The first difficulties came at the end of the 1950s with a fall in prices. The aperitif producer Bartissol had cellars in Banyuls, and when they went bankrupt all their stock was sold to Mas Amiel. In the twentieth century Banyuls and Collioure remained almost entirely ignored outside their own region, and that situation largely continues today.

Vincent talked about the *bail à complant*, a system of land rental which dates back to the Middle Ages. There are three signatories; the buyer, the vendor and the owner of the land. What you plant on the land is yours, and there is no fixed term so your children can inherit; your rent is a percentage of the grapes to the owner of the land, who these days is often a cooperative member. Apparently, this method of running the land was encouraged after the region became part of France, in an attempt to discourage smuggling. The would-be smugglers were encouraged to become vignerons instead!

The total area of the two appellations currently stands at a little over 1,300 hectares, and is well-nigh impossible to increase. On the contrary, the tendency is for the vineyard area to fall. The Spanish frontier limits the appellation to the south and the northern edge is where the hills fall into the plain and the schist ends; to the west there are mountains, and the Mediterranean is to the east. The soil is schist, through which the vine roots filter to find water, and many of the vineyards have drainage systems, locally called *agulles*, or drainage ditches, designed to limit the damage from the occasionally excessive rainfall. When it rains, the force of the water can be violent, with rivers bursting their banks. The average annual rainfall is about 900 millimetres, but that figure has become much less consistent in recent years.

The terraced vineyards are virtually impossible to work mechanically. Wider terraces will accommodate a small tractor; weedkiller is now rare, and these days they may till, or leave some grass to help retain the soil, as a measure against erosion. Pierre Gaillard of Domaine Madeloc, who knows both the Côte Rôtie and Banyuls, considers the slopes of Côte Rôtie to be steeper and more challenging. Here the terraces are quite regular, whereas Côte Rôtie has even narrower terraces. Côte Rôtie also has more regular rain, and the vineyards have some soil, whereas here it is solid rock. Emmanuel Cazes of Les Clos de Paulilles contrasted the difficulties of Banyuls and Rivesaltes. In Banyuls it takes them ten days to treat 90 hectares but in their flatter vineyards of Rivesaltes, they can treat 180 hectares in 36 hours. However, there is very little mildew in Banyuls with so many vineyards by the sea, while those further away from the sea are more sensitive to disease. For Laetitia Pietri-Clara of Domaine Pietri-Géraud, the greatest problem is the wind. If the vines need support they will be planted *en echalas*, with a single supporting post; otherwise they are *gobelet* or bush vines. A *palissage* of wires is impossible. The vineyards are difficult to work and the terraces are essential, but not all are planted. Indeed, the vineyard area is declining annually, especially amongst the cooperative members, with the next generation not wishing to continue their parents' work.

It is the proximity to both the sea and the mountains that determines the character of Banyuls and Collioure, with a multiplicity of sites, terroirs and grape varieties. You will find wines with a maritime influence, while others enjoy the benefits of altitude and the proximity to the mountains. The differences are exemplified in Domaine de la Rectorie's two cuvées, Côté Mer and Côté Montagne. It is this dichotomy that distinguishes Banyuls from the other Vins Doux Naturels and Collioure from the other vins secs of Roussillon. The terroir allows for enormous diversity, with differences of aspect and altitude. The winemaking is no different than for the other wines of Roussillon, with each wine grower having their own particular ideas and philosophies.

The syndicat of Banyuls comprises three cooperatives and a total of 54 estates, of which 15 are significant producers. The three cooperatives account for 80 per cent of production. Some of the independent wine growers criticize the cooperatives, or feel overwhelmed by them, saying they dominate the appellations in volume and complicate matters. For financial reasons, they need to sell their wines quickly, and that is not always good for quality. Michael Berta ruefully observed that when the cooperative is 'ill', referring to the large Terres des Templiers, 'we all catch a cold'.

The largest producer of all, created in 1950, is a group of five cooperatives, the Groupement Interproducteurs Collioure Banyuls, which includes the cooperative of Collioure, the Cellier Dominicain and is now called Terres des Templiers. The Cave l'Etoile, also in Banyuls, stands apart. There is no doubt that more interesting wines come from the independent wine growers. The Parcé family, as notable vineyard owners, have played an important role in the development of Banyuls and there are other families with a long history in the area, not only as wine growers but often as fishermen too.

Vincent Parcé made some pertinent observations about the shift from sweet to dry wine. The cooperatives made wine in volume when people really did not know how to make vin sec, since sugar can camouflage many defects, he explained. The image of Collioure was of robust, powerful, hard red wines, but made with care and talent Collioure can be fresh and elegant. Winemaking has improved here with the modern developments common to winemaking all over the south of France, and so the image is improving too. For Vincent, white Collioure is more identifiable, and more distinguishable, whereas red Collioure is less obvious, with so many different expressions, depending on the precise blend of Syrah, Grenache, Carignan and Mourvèdre. He described Collioure as the Côte Rôtie of Roussillon, although Côte Rôtie, with 200 hectares, is much smaller.

Once the wines were for volume consumption, but that has changed as fashions change. Emmanuel Cazes suggested red Collioure as an alternative to Châteauneuf-du-Pape. For wine producers the crucial difference between Collioure and the more famous appellation is that the wines need to be explained before they can be sold. If Collioure is unknown, Banyuls is unloved, but, says Emmanuel, 'Banyuls is our history and it has led people to vin sec; but they should consider the originality of the Vins Doux Naturels, and encourage the *cavistes* or wine shops to offer them for tasting. Once they try them, people love them. But if they are too expensive, they will not sell,' he says.

Banyuls has no colour specified in the appellation decree. All three colours of grapes are *complanté* so you can have Grenache Noir, Grenache Gris and Grenache Blanc in the same vineyard. The only constraint before the Second World War was a minimum of 50 per cent Grenache Noir; the colour of your wine depended on how you vinified your grapes, and in the 1950s Banyuls tended to be lighter in colour. Vincent commented that his grandfather and father both liked oxidation; that was what mattered. Originally, they made Vin Doux Naturel and Rancio Sec – Collioure was originally a Rancio Sec. Not so long ago, the production of vin sec was tiny compared to that of Vin Doux Naturel, but that has changed dramatically, especially since the creation of the appellation of Collioure.

Essentially there are two styles of Banyuls, reductive and oxidative. Rimage is a relatively new creation, a young vintage wine, bottled after 18 months, with no oxidative ageing. The appeal is immediate. Michel Berta described it as a *pied de nez*, a snub to vintage Port. Rime means *raisin*, or grape, in Catalan, so rimage literally means grapes with an age or vintage. Another interpretation of rimage is the *vin de l'année*, or wine of the year. Essentially it is made like a Collioure, apart from the *mutage*. Elise Gaillard at Domaine Madeloc destalks her grapes to avoid any bitterness and then gives them a five-day cool pre-fermentation maceration. The time on skins for Rimage is three weeks, whereas it may be as long as six weeks for Collioure.

The oxidative styles include Traditionnel, which may be non-vintage, aged for a minimum of three years, and Hors d'Age, aged for a minimum of five years. Often it may be much longer, and entail a solera. The élevage may be in barrels, of varying sizes, including large foudres, which are not topped up. The small barrels may be in a cellar, or in an attic, or even outside, exposed to the elements. The larger the surface of wine in contact with air, the more evaporation there is. Hygiene is paramount if you are using old wood. The other ageing vessel is the glass *bonbonne*, where the wine might remain for a year or so, again outside and exposed to the elements and changes in temperature.

Banyuls Blanc is generally reductive and designed to be drunk when it is young and fresh. The grape varieties are Macabeo, Grenache Blanc and Grenache Gris – Muscat rarely features in the vineyards of Banyuls. Tourbat is a traditional white variety with good natural acidity. Pierre Gaillard has replanted some, observing that it is good for Vin Doux Naturel, but not for vin sec. An Ambré is an Hors d'Age which began life as Banyuls Blanc, but turned amber with age.

The category of *grand cru* is peculiar to Banyuls. It was recognized as a separate appellation in 1962 and requires a minimum of 75 per cent Grenache Noir, in practice usually almost pure Grenache Noir. It is a vintage wine with a minimum of three years of ageing, without any oxidation, and the alcohol level should be 22.5% abv. When a *grand cru* is more than five years old it can be described as an Hors d'Age, and indeed *rancio*, with the appropriate characteristics, except that most *grands crus* do tend to be reductive, while Hors d'Age wines are firmly oxidative. There are mixed reactions to the term *grand cru*; some love it, others hate it. Emmanuel Cazes suggested that a clarification was needed; they make a reductive *grand cru* with the barrels regularly topped up so that the wine tastes like a Rimage, but with more depth.

Laetitia Pietri-Clara talked about the image of the *grand cru*. Essentially it is a Rimage, as it is a vintage wine, for oxidation destroys the particular characteristics of the vintage. For Laetitia, it should only be made in the best years and from the best plots. Vincent Cantié does not believe in the *grand cru*. He sees it as a commercial tool rather than a recognition of quality. 'It is what we do anyway, and does not correspond to a commercial reality. There is no real difference of expression. Elsewhere a *grand cru* is the result of site selection, not a difference in winemaking,' he says.

There is no doubt that the market for Banyuls is far from easy. The wines no longer enjoy their previous popularity, vignerons have to work harder to earn money from Banyuls than they do from vin sec. The roots of the word *mutage* are in Spanish, Italian and Catalan, not French (Arnaud de Villeneuve, who discovered the process, was Catalan). It literally means to render the wine *mute*, or silent, in other words conjuring up an image of silencing the *esprit*, the spirit or alcohol of the wine.

Philippe Gard of Coume del Mas would like to make Vin Doux Naturel with a lower alcohol level and with only 40–50 grams per litre residual sugar. Pierre Gaillard enthused about Vin Doux Naturel. He loves it, but the market is complicated, so consequently three quarters of his production is Collioure, and two-thirds of that is red. Laetitia described Banyuls as a niche wine; it still accounts for 40 per cent of her production, while she considers Collioure to be less well-known than Banyuls. Vincent Cantié says that sugar was once the aristocracy of wine, but no longer. The Vin Doux Naturel market has collapsed. Michel Berta also talked dispassionately about the problems of Banyuls, saying that France knows the name, but does not know what it is worth. There is no recognition; the vineyard area is declining, in danger of falling below 1,000 hectares, when once it totalled 3,000. 'The French do not know how to drink Banyuls; they think of it as an aperitif, treating it like Port, which in France is virtually all drunk before a meal. However, the alcohol in Banyuls is too strong for a pre-dinner drink. Often it is badly served, and also badly stored. And yet it is a great classic – we are there to make people dream, and we have everything to enable them to do so,' says Michel. When asked about projects for the future, he ruefully said, 'to survive. Times are hard. We have a history of adapting, but there are problems: the impossibility of mechanizing the vineyards, global warming, with an increased frequency of drought, and with that, lower yields and yet another problem.' He would irrigate if it would save the vines, but the terrain rarely makes that practicable.

Olivier Sapéras of Domaine Vial-Magnères also has a dispassionate view of the future. The small estates are no longer viable and as a scientist, he also has regular employment in an agricultural laboratory in Perpignan. Production costs are high, and with the recent drought years provoked by climate change, they are all making less wine for the same production costs. Everybody is pretty much in the same boat. As a *propriétaire récoltant*, you cannot be a négociant as well, so there is no logical way of increasing your production unless you rent vineyards, but that is not necessarily economically viable. Prices too have fallen, and that is a problem. For example, a delicious 2007 Banyuls *grand cru* from Olivier retailed at €34 in 2019 and you simply cannot earn a living with wine at that price. Until they find a way to add value to their wines, as other fortified wine producers such as Port have, this will continue to be a problem. However, Olivier was remarkably sanguine, commenting that his father always said the best vintage was the one that is coming.

Domaine Augustin with Domaine de la Préceptorie

Banyuls-sur-Mer and St Paul-de-Fenouillet

www.domaineaugustin.com / www.parce-freres.com

Vincent Parcé is a member of the Parcé family, which did so much to establish the reputation of Banyuls. The legendary Dr André Parcé was a cousin of his grandfather, while his father, Marc, and his brother Thierry created Domaine de la Préceptorie, which is based in St Paul-de-Fenouillet. Vincent's brother Martin is responsible for the management of that estate, while another brother, Augustin, makes the wines for both Domaine Augustin and Domaine de la Préceptorie. As I tasted the wines of both estates in Banyuls, it makes sense to write about them together. Domaine Augustin comprises eight hectares, in about 13 plots, including what Vincent called the historic vineyard, which came from his great-great-grandmother, Thérèse Puig, who apparently was another of the great characters of Banyuls. These days they produce 90 per cent Collioure and just 10 per cent Banyuls. The wines are made in a cellar in the village of St André near Argelès.

In addition, there is a négociant business, Les Vins Parcé Frères, which enables them to increase their production with the purchase of grapes. Vincent's responsibilities are commercial and he has a perceptive view of the problems of selling Banyuls and Collioure.

We met in the substantial town house in the avenue Puig del Mas from which Vincent runs the business. It was built in the eighteenth century and still has the original floor tiles. We sat at a large table in the kitchen and tasted the wines of the two estates. Vincent considered the differences between Collioure and Maury. In Collioure you have the influence of the sea, and the wines are more elegant, whereas Maury is sunnier and much hotter so the wines have more breadth. Côtes du Roussillon Coume Marie Blanc, from La Préceptorie, is a blend of equal parts Grenache Gris and Grenache Blanc, with Macabeo, some *complanté*, from several plots between Estagel and St Paul-de-Fenouillet. Along with the red Coume Marie, this is the heart of the estate, which comprises 29 hectares in 44 plots.The white is fermented in barrel, with good texture and balancing acidity; the red is a blend of Grenache with Syrah, Carignan and Mourvèdre, aged in both wood and vat, with some spicy red fruit.

The first Collioure in their range, mostly from Grenache with some Carignan and just 5 per cent Mourvèdre, aged mainly in barrique, was ripe with some fresh spice. Adéodat, after the son of St Augustin, and meaning *ce que Dieu donne*, what God gives, consists of Grenache, with possibly as much as 20 per cent Carignan in an old field blend, from a 2.5-hectare plot just outside Banyuls. Adéodat Blanc, a pure Grenache Gris, is fermented and aged in wood, making for texture with ripe, floral fruit and balancing acidity. Maury Sec Domaine la Préceptorie SO2 Mon Amour, comes from an old plot of Grenache and Carignan, with some elegant fruit. Vincent admitted that the name is deliberately provocative, observing that natural wine should have a drink by date as it is a perishable product.

We continued with Vins Doux Naturels. Banyuls Rimage, a pure Grenache Noir, *muté sur grains* and kept in vat, was redolent of cherry liqueur, with a balance of sugar, acidity and tannin. Maury Cuvée Aurélie from la Préceptorie comes from young vines, with ripe fruit and spice, and even more liqueur cherries. After 36 years in barrel, without any *ouillage*, a delicious 1982 Banyuls had a firm, nutty nose and palate.

Our tasting finished with a Vin de France, Vinum, from La Préceptorie, from Grenache Gris, Grenache Blanc and Macabeo. The wine has been left outside to develop in barrel, without any *ouillage*, so that it might or might not develop a *voile*. It was quite delicious – light amber in colour, and on the palate, salty and incisive with firm acidity.

Domaine Berta-Maillol

Banyuls-sur-Mer

www.bertamaillol.com

Michel Berta is descended from Marie Maillol, sister of the sculptor Aristide Maillol. There is a document, dated 1611, attesting the existence of the family estate, with an inventory of the cellar, and the purchase of an alembic. Michel's great-grandfather was one of the first in the region to put wine in bottle, before the First World War when it was easier to transport wine in barrel and most wine was sold through the *négoce*. The large house on the seafront, the Hôtel Maillol, is a witness to the family's past glory but was sold after his grandfather's death, although they retained the adjoining shop, where we tasted. Michel's father, Yvon, continues to be involved with the Maillol association, with Michel observing that the locals do not know much about the sculptor. For me the artist's graceful figures on the seafront add to the charm of Banyuls. Michel joined the family company in 2003, working with his father and brother, Jean-Louis, who runs the vineyards. They make the wine together, with their father, a lively octogenarian, supervising.

White Collioure La Rivière is a pure Grenache Blanc, given some skin contact, but no malolactic fermentation. It was rounded, with a citrus note and some firm tension. Michel pointed out that their rosé, La Tina, does not conform to current fashion. They wanted a wine that would keep, and go well with food. It was satisfyingly vinous, from Grenache and Syrah, and *saignée* rather than pressed. Arrels, meaning *racines* or roots in Catalan, is a blend of Grenache, Mourvèdre and Carignan aged in vat. Barral, which is Syrah and Grenache aged in barrel, has some powerful fruit, contrasting with the freshness of Arrels.

Next came a range of Banyuls. A Blanc, from Grenache Blanc and old vines of Grenache Gris with some Macabeo, was rounded and honeyed with notes of aniseed and almond. The Rimage was ripe and spicy and aged only in vat, otherwise it would lose its essential freshness. This is the base of all their Vins Doux Naturels. They do not make a *grand cru*, but do produce an oxidized Banyuls Tradition, with sweet, rich fruit and notes of prunes. A drier 2002 was wonderfully nutty. 'A great classic,' observed Michel, lamenting that people do not know when to drink these wines. Hors d'Age came from a solera with wines that are at least ten years old. It was dry, nutty and quite delicious.

Coume del Mas

Banyuls-sur-Mer

www.coumedelmas.fr

Coume del Mas has its cellar up in the hills, just outside the village of Cosprons, with a wonderful view of the sea and the terraced vineyards. Philippe Gard began by introducing the estate, explaining that he had been brought up in Paris and had worked in Bordeaux and Burgundy before coming to Banyuls. His mother was from Sancerre and he has cousins with a wine estate there. However, his grandparents came from Cantal and his great-grandfather had vines in the little known Aveyron appellation of Entraygues-le-Fel, where his daughter, Fanny, is now producing some Chenin Blanc and Pinot Noir. Perhaps amusingly she is qualified as a water engineer. Philippe commented that 100 years ago that appellation was like Banyuls, with lots of tiny plots on steep hillsides. His first vintage at Coume del Mas was 1997. He has 12 hectares divided into about 35 plots, in three different zones of Banyuls. His aim is to stay small, to retain *la taille humaine*.

Philippe is also associated with Mas Cristine, situated between Collioure and Argelès, working with Julien Grill. They took over the estate from the Dauré family in 2006 and it complements the range, with different soils, but the same philosophy. However, Philippe observed that Mas Cristine is 'easier and more technical'. Andy Cook, Philippe's colleague, arrived from New Zealand in 2007 to do a *stage* at Coume del Mas and stayed on, now making the wine at Mas Cristine. They also have Terrimbo, a small organic estate in Collioure. Altogether they make about 30 different wines, including *vins de cépages* at Mas Cristine, and also have a small négociant activity called Tramontane for varietal *vins de pays*, Macabeo and Grenache Noir.

Tasting with Philippe is fun; he is lively with plenty of ideas and a perceptive view of Banyuls and Collioure. In his winemaking he is fairly *laissez-faire*, using natural yeast and allowing the white wines to do a malolactic fermentation, if that is what happens. There were numerous highlights from my most recent tasting. Folio, mainly from Grenache Gris, was vinified in old wood and has depth and salinity. Folio Edition Spéciale, from Grenache Gris grown on black schist, with a proportion aged in new wood, is made 'more in the spirit of Burgundy', according to Philippe. Terrimbo, from Syrah and Grenache grown close to the sea, had some peppery fruit from the Syrah, with more elegance than Schists, a pure Grenache. Quadrator, from Grenache with 20 per cent Carignan and 30 per cent Mourvèdre, grown on schist, is the heart of the range of Coume del Mas and a rich, powerful wine. Abyss is equal parts Syrah and Grenache, given a *vinification intégrale*, to produce a very perfumed and intense wine with tapenade fruit. In some instances, we compared the 2017 with the 2018 vintage, with Philippe commenting that in 2018 there was a greater variety between the different terroirs, whereas with the heat and sunshine of 2017, everything was more uniform. The years 2018 and 2019 were challenging for the winemaker, and much more exciting

Galateo Rimage followed, with some fresh spice, but not too much sugar. Philippe mutes at 80 grams per litre residual sugar, so that the tannin balances the sugar, but would like to be able to mute at even lower sugar levels. Quintessence is a pure Grenache Noir *muté sur grains*, with a long extraction and considerable concentration balanced with a tannic structure. It is not made every year; Philippe considers that Galateo provides an introduction to Banyuls, whereas Quintessence is for connoisseurs. An Hors d'Age, bottled after five years, was firm and nutty; the barrels are kept above the vats, under the roof. Another Ambré Hors d'Age spent ten years in barrel and was firm and nutty, with salty, incisive notes. Philippe commented that his parents did not drink Vin Doux Naturel so it was a long road from Sancerre to Ambré. He is passionately committed to the region and its wines.

For an account of the harvest at Coume del Mas, with Philippe and Andy, I can recommend Richard Bray's *Salt and Old Vines*. He conveys the enthusiasm and fun, and also the immense hard work. Richard worked his first harvest at Coume del Mas in 2008, and now makes his own small batches of wine, under the label Cathar(tic) Wine. I tasted his wines in London, starting with a 2019 white called Neuks 'n' Crannies. A blend of Roussanne, Carignan Gris and Grenache Gris, it is given some oak ageing and has some rounded, creamy fruit. The acidity of the Carignan balances the ripe fruit of the Roussanne. The 2019 rosé, By Any Other Name, a pure Mourvèdre pressed very gently and racked into a barrel with some lees, is a food rosé with elegant weight and some ageing potential. The red wine, Sriracha, is pure Syrah and remarkably peppery for a pure Syrah from Roussillon, with a satisfying freshness on the finish. Richard explained that the hot, Asian sauce sriracha is the winery's favourite condiment – he recommends it with a Catalan sausage. The label Cathar(tic) Wines refers to the history of the Cathars, and also reflects the cathartic experience of winemaking.

Philippe has also encouraged his young oenologist Léah Anglès to produce her own wines, first a red, a pure Grenache, and then a white from Grenache Gris and Grenache Blanc, under the label, Divay.

Domaine Bruno Duchêne at Les 9 Caves

Banyuls-sur-Mer

www.9caves.com

Bruno Duchêne bucks the traditions of Banyuls. He took over some of the old cellars of the Templiers cooperative, which had been used for storage. At the front, there is a restaurant offering creative dishes and an imaginative wine list, run by a Dutch couple since 2017. I can recommend its lively atmosphere and delicious tapas. On the first floor are some gîtes and at the back of the building there are cellar facilities for nine small growers, including Bruno. They are all organic and all make natural wine, amounting to about 50 hectares of vineyards. Bruno explained that he came to Banyuls in 2002 and bought some vineyards: two hectares became four hectares, both close to the sea and up in the mountains, with lots of old vines. He also buys grapes from friends and altogether vinifies about 15 hectares.

A very genial and eclectic tasting followed with Bruno saying, 'let's try this', as we went from vat to barrel and back to vat in his cellar. First there was a rosé, a blend of Syrah, Mourvèdre and Grenache, that had just finished fermenting. La Luna Blanc is a Côte Vermeille as it includes Chardonnay as well as Vermentino and is sold on 1 March, whereas Collioure cannot be sold until 1 June. Vall Pompo white comes from Grenache Blanc, fermented in barrel. Red La Luna comes from all three colours of Grenache, plus a little Carignan and Mourvèdre, and accounts for half of Bruno's production; it was rounded and perfumed. La Pascole comprises two plots, field blends of old Carignan and Grenache, one east facing at 150 metres and the other south facing at 50 metres. It had perfumed red fruit and an elegant fresh finish.

Bruno talked about his earlier career, explaining that he comes from the Loire Valley and had worked with mushrooms before learning about viticulture in Burgundy. He displays a friendly balance of humour and energy. He enthused about amphorae, with Inés, a Macabeo kept in an amphora. There is more oxygenation with amphorae than with wood, but you lose less wine. The difference between Banyuls in amphorae and in barrel is very marked. Rimage, kept in barrel and muted when the sugar was at 70 grams per litre, had fresh perfumed red fruit. A Banyuls from amphorae, from two different plots in Cerbère, was perfumed and concentrated. Finally, there was a Naturellement Doux Vin de France, without any *mutage* which was more intense, with more sucrosity. I sense that things change with every vintage, as the mood takes Bruno, and that of course is part of the fun.

In addition, he collects grapes for more winemaking after his own harvest has finished, from as far as way as the Côte Chalonnaise, Montlouis in the Loire Valley and Priorat in southern Catalonia, and brings them back to Banyuls in a lorry. It's a weekend trip away.

Cave l'Etoile

Banyuls-sur-Mer

www.banyuls-etoile.com

Bruno Cazes has been the director of the emblematic cooperative of Banyuls, L'Etoile since 2008. He is a member of the Cazes family in Rivesaltes but left the family business to seek a new challenge when his father, André, retired. Bruno admitted that it was a shock when he started, entailing a journey back in time and some drastic reorganization, for the previous director had been there for 35 years. The first thing to do was an inventory, checking all the old barrels, which took two years. Bruno felt passionately that these were wines that must be saved before they disappeared and essentially, he has created a unique collection of Banyuls. They are the only cellar to have every vintage since 1947, no more than 5 hectolitres of any one wine, in *bonbonne*, barrel or bottle. It is a veritable vinous treasure and Bruno is justifiably proud of what he has achieved.

Cave l'Etoile is not associated with the other cooperatives of Banyuls. The official date of foundation is 1921, but the cellar existed before then. Created by just 12 wine growers, it now has 100 members, some with very tiny plots. Twenty per cent of the members account for 80 per cent of the volume.

We wandered round the cellar, which was built in the late nineteenth century. The smell of mature wine was intoxicating. The tasting took place in the shop and began with a floral Collioure Blanc. Collioure Rouge Montagne, with no wood, was fresh with red fruit, while Le Clos de Fourat, from a vineyard near Paulilles, had good structure. Banyuls Blanc was lightly honeyed, with notes of *crème pâtissière*. Rimage had ripe berry fruit, while an Hors d'Age was elegantly nutty. My tasting finished with 1992 Select Vieux *grand cru*, with a deep mahogany colour, a dry, nutty palate and intense notes of coffee and chocolate. It had been aged partly in foudres and partly in demi-muids.

Although Bruno is passionate about old Banyuls, with oxidative Banyuls the tradition of the appellation, he is realistic enough to believe that it is Rimage that will save the appellation. With its fresh fruit, it is the wine for the new generation. They have also had some success with Banyuls Blanc and have been making Banyuls Rosé since 2011. They continue to make more Vin Doux Naturel than vin sec, 60 per cent to 40 per cent, quite unlike most other producers in the appellation.

Domaine Madeloc

Banyuls-sur-Mer

www.domainespierregaillard.com

Pierre Gaillard had already established a reputation as a talented producer of Côte Rôtie when he decided to buy vineyards in Banyuls. That begs the question: why? He chuckled; he was seduced by the scenery, with slopes that are as steep as those of Côte Rôtie. He talked about the beauty of the land, the sun, the schist, the landscape. He enthused about Grenache, describing it as the Pinot Noir of the south, admitting that he loves Pinot Noir, but can't afford it, so he came here in 2002 and bought vines. It was the result of a chance meeting with Marc Parcé at a wine fair, who told him about an opportunity, some vines that were for sale. He sensed the huge viticultural potential of these 17 hectares across 20 plots, both by the sea and at altitude. The former owner, an old lady, had virtually abandoned them; her last crop had been precisely 42 hectolitres. Pierre replanted and has reduced the holding to ten hectares.

His daughter, Elise, now runs the estate. She has been living in Banyuls since 2009 and now proudly calls herself a *banyulencque*. As she says, she gave herself two years, and forgot to go home. Her father gives her a free rein and from time to time 'comes to build some drystone walls'. Elise studied in Angers and has done *stages* all over the world, in Chile, South Africa, Italy and Spain. Tasting with father and daughter in their cellar in the centre of the town was great fun. You sense a good relationship and a shared sense of humour. Collioure Blanc Cuvée Tremadoc is a blend of Grenache Gris with some Vermentino and Roussanne, with some oak ageing in old barrels from Condrieu and some *bâtonnage*. They had not intended to make white wine, but Grenache Gris does so well here, while Vermentino adds acidity. They planted Roussanne for its association with the Rhône Valley, but do not find it as successful as they had anticipated. It was too bitter and simply did not excite. Collioure Blanc Penya, from Grenache Gris and Vermentino, is a selection of barrels, with more structure and incisive acidity.

We went on to red wines, with Cuvée Serral, which is based on Grenache with some Carignan and Mourvèdre and had fresh, leathery fruit. Cuvée Magenca, from Grenache, with some Carignan as well as Mourvèdre, was deep in colour, with smoky, structured fruit, good acidity and an elegant finish. Mourvèdre copes well with wood and also has great potential. Collioure Cuvée Crestall, Mourvèdre with a theoretical percentage of Syrah, was fresh and firm with balance and potential. Pierre commented that at the beginning he was much more experimental, making several different cuvées, so that he could really understand the terroir. Pierre considers the main defect of Grenache is its alcohol: it must have body to counterbalance the alcohol, and he refuses to pick his grapes too early; they must be ripe. Carignan was the surprise and he would like to plant some more. He finds that it gives higher yields than Grenache Noir. It was planted in richer soils at the bottom of slopes. However, a vineyard solely of Carignan does not exist as it was always *complanté* with Grenache, usually accounting for about a tenth of a vineyard. The addition of Mourvèdre gives spice.

Banyuls Blanc Asphodères was rounded and honeyed, coming from Grenache Gris, as well as Tourbat, which adds acidity. All good Banyuls is made by blending; they started by experimenting, trying different ways of oxidation and also more modern wines without any oxidation. The image of Banyuls Blanc is rather old fashioned. The Rimage, Cirera, which means cherry in Catalan, was muted at 90 grams per litre residual sugar and kept in 30-hectolitre foudres that are topped up, with an emphasis on the fruit. A non-vintage Banyuls Robert Pagès is a blend of two vintages. A proportion spends six years in *bonbonnes*, five years inside and one year outside, some spent time in foudres (not full) and the rest was in barriques outside. It was rounded, with nutty fruit and lots of nuances. With old wood, you must take care with the hygiene to retain the purity of the fruit. Our tasting finished with a Solera Hors d'Age. The solera was begun in 1920, forgotten in 1962 and rediscovered in 2002. The wine was intense, reminding me of an old oloroso, with extraordinary concentration.

Les Clos de Paulilles

Port-Vendres

www.lesclosdepaulilles.com

Les Clos de Paulilles is a stunningly dramatic spot right on the coast between Collioure and Port-Vendres. Even on a grey day the vineyards were wonderfully atmospheric. They enjoy the influence of the sea, with the Grenache on steep slopes and Carignan and Mourvèdre at the bottom. As we walked round the vineyards, Emmanuel Cazes talked about the history. Essentially there are three beaches, and during the Franco-Prussian War of 1870, the French government decided to build an armaments factory as far away from the frontline as possible, and this is the spot they chose. President Léon Gambetta inaugurated the factory after the Franco-Prussian war had finished, and it supplied dynamite for the construction of the Panama Canal. In 1920 the Nobel family took over and it became a closed community with people from the French colonies working and living there. It ceased to function in 1970 and then in 1980 the *loi littoral* was passed, to protect the coastline. Most of the factory buildings were destroyed, and as the area had been very polluted, vines were the only possible activity, certainly not tourism.

The estate was developed by Bernard Dauré of Château de Jau, who replanted most of the vineyards, some 90 hectares out of 110. He sold it to the Cazes family who made their first vintage in 2013. The soil is brown schist. Weedkiller was used in the 1960s but now they are working the soil. Erosion is a problem. The site is particularly good for white vines, so they are replanting Grenache Blanc and Grenache Gris, but not Macabeo or Tourbat. They may include a little Vermentino. All the vines apart from the Syrah are bush vines, with quite short branches, because of the wind. Emmanuel talked about the differences between the vineyards here and their vineyards in Rivesaltes. Here only a third of the vineyards are mechanizable; the remaining area is much more complicated. In the 90 hectares, there are 130 different plots, and they do a lot of site selection.

The old cellars have thick walls made from local stone. There are concrete vats as well as stainless steel and some concrete eggs, and they work by gravity. The traditional *bonbonnes* were apparently originally used for transporting the sulphuric acid for the dynamite. They have an attractive restaurant, where we tasted. Collioure Blanc is based on Grenache Blanc with just 5 per cent Grenache Gris. It is made without any wood as they consider that wood masks the fruit and want to retain the salinity and iodine notes of the wine, with a firm, stony freshness. Cap Béar Blanc comes from the vineyard closest to Cap Béar itself. The wine is pure Grenache Gris, with some oak ageing and also élevage in an egg, making for an appealing richness, almost unctuousness, balanced with a streak of tannin. For Emmanuel, it proves the potential of the vineyards for white wine. Cap Béar Rosé is a blend of Grenache Noir and Syrah with some ageing in oak, making for some weight and texture with firm fruit. Their red Collioure is based on Grenache with some Mourvèdre, Syrah and Carignan; aged in vat, it was firm and gutsy. Emmanuel says there are two styles of Collioure – modern and classic – and these are modern. Cap Béar Red is mainly Syrah with a little Grenache, with one-third aged in oak, so that it is firm and peppery, with smoky notes from the schist.

Rimage was muted on the skins, with some ripe cherry fruit. Banyuls Tradition was aged in *bonbonnes* and then blended in vat, with some red fruit, notes of oxidation and a tannic streak. Banyuls Blanc comes from Grenache Blanc, muted *sur jus*, with notes of aniseed. Emmanuel talked amusingly about what he called the war between Collioure and Banyuls, while the Cazes family are considered foreigners as they come from Rivesaltes.

Domaine Pic Joan

Banyuls-sur-Mer

www.domaine-pic-joan.fr

Jean Solé and his wife, Laura Parcé, have a smart wine shop on the hill just outside Banyuls, although they were busy mopping up flood water when we arrived, as it had rained so hard the previous night. Jean has always worked with wine, taking his grapes to the cooperative, which he left in 2009 to create his own estate, producing his first serious vintage in 2010. Meanwhile Laura's father, Jean-Michel Parcé of Domaine du Mas Blanc, was retiring, and so they took over his cellars in the centre of Banyuls, and incorporated Domaine du Mas Blanc into Domaine Pic Joan, with 14 hectares in 15 small plots, all in Banyuls.

They make an extensive range of wine, including an entry level Collioure Insolence in red and white, Collioure Pic Joan in all three colours and, for Banyuls, a Rimage and a white, as well as a Traditional Banyuls Hors d'Age, aged in both barrel and in *bonbonnes*. However, they had sold out of quite a few wines so our tasting was quite limited. The white Collioure comes mainly from Grenache Gris with some Grenache Blanc, fermented and aged in wood for six months, with some *bâtonnage*, but no malolactic fermentation. It was floral and elegant, with well-integrated oak. In contrast Insolence Blanc is only kept in vat, with more citrus fruit. Insolence Rouge is a blend of two-thirds Grenache Noir and one-third Syrah, kept in vat, with some ripe, spicy fruit, intended for easy drinking. Pic Joan with Grenache, Carignan and some Mourvèdre, kept in barrel, was firm, peppery and structured, with nicely integrated oak. Rimage came next, muted at 80–90 grams per litre residual sugar as Jean does not like Banyuls with too much sugar, which makes it too heavy. This was ripe and spicy. There are projects for the future and they deserve to do well.

Domaine Pietri-Géraud

Collioure

www.domaine-pietri-geraud.com

Laetitia Pietri-Clara is the fifth generation of wine growers from her family in Collioure. Her great-grandparents and grandparents were doctors as well as vignerons. As she explained, everyone in Collioure had some vines, and most people were also fishermen. Her grandfather was one of the first to have his own press and people brought their grapes to him. He helped found Le Cellier des Dominicains, but did not actually join the cooperative. Her mother, who married a M. Petri, took over her father's estate, but as Laetitia observed, he had not taught her mother anything, for he had not anticipated the arrival of the vins secs, or indeed the development of modern oenology. Laetitia is the first member of her family actually to earn a living from her vines, making her first vintage in 1997.

Laetitia is bright and vivacious and has a welcoming shop in the centre of Collioure, complete with an old armchair from her great-grandfather's surgery, and a compact cellar in another street close by. It is not easy working in the centre of the town, with the constraints of space that entails. Laetitia explained that the first challenge was to renovate, and modernize, to replant much of the vineyard and make the estate financially viable. She has some 16.5 hectares altogether, about 25 plots in 10 blocks, nearly 13.5 hectares of Collioure and 3.5 of Banyuls.

Our tasting began with Collioure Rosé Le rosé de mon père, as her father loved rosé. It is a blend of Grenache Noir, Syrah and Mourvèdre, fresh and rounded with a salty note, and light in colour 'following the dictates of fashion'. Collioure Rosé L'Ecume comes from a field blend of Grenache Gris, with a little Grenache Blanc as well as some Vermentino and Marsanne, planted by her mother, initially on an experimental basis. The wine is aged in 400-litre barrels and was elegantly oaky.

Collioure Rouge Sine Nomine is based on Grenache, with some Syrah, a little Mourvèdre and some Carignan planted by her great-grandfather. Kept in vat, it is ripe, *gouleyant* and spicy, and with the four varieties, very characteristic of Collioure. Collioure Le Moulin de la Cortine, based on Syrah grown at altitude, with some Grenache and Mourvèdre, has firm, peppery fruit and more structure. Trousse Chemise is based on Mourvèdre, which Laetitia considers to be a more complicated variety. It is given a *vinification intégrale* and is blended with some Grenache so that it is ripe and intense.

Laetitia was one of the first to make a white Banyuls in 1992. Her Banyuls Blanc from Grenache Blanc spends a year in oak and is the perfect accompaniment to a *crème catalane*. Her Muscat de Rivesaltes comes from a vineyard near Elne. Banyuls Rosé is fun and fruity. 'It must not be taken seriously,' says Laetitia, describing it as 'Banyuls *décomplexé*'. Cuvée Rimage Mademoiselle O, for her daughter, Ornella, has ripe fruit, and a Banyuls Tradition, Cuvée Joseph Géraud, named after her grandfather, is a blend of Grenache Noir and old Carignan, aged for seven years in foudres, which are never full. It was nicely spicy and long. Cuvée Mediterranée is a pure Grenache Noir that spends five years in barrel, with some *ouillage* to limit the oxidation, so that it is fruitier. Laetitia refuses to make a *grand cru* as she feels it is a poorly defined term. Cuvée du Soleil is a Banyuls Ambré, aged outside for four years, which induces a lot of evaporation so that the flavours are firm and nutty. Our tasting finished with the Hors d'Age, from a solera started by her great-grandfather; the average age of the wine is 20 to 25 years and the flavours are concentrated and nutty.

Domaine de la Rectorie

Banyuls-sur-Mer

www.rectorie.com

Domaine de la Rectorie is run from a large town house in the centre of Banyuls, where I spent a convivial hour or so tasting and chatting with Jean-Emmanuel Parcé and his Argentinian winemaker, Emilio Perez. Jean-Emmanuel is a member of the Parcé family who did so much to develop Banyuls and Collioure. He explained that the estate was created by his uncle, Marc, and father, Thierry, in 1984. It has grown gradually, and now covers 35 hectares in 46 plots in Banyuls, Cosprons and Port-Vendres. Marc is now involved with Domaine Augustin, and Jean-Emmanuel started working in the business in 2010, after studying viticulture and winemaking at Carcassonne. He has also developed an estate in Aniane in the Hérault, Domaine la Chapelle St Mathieu, planting 10 hectares of cleared *garrigue*. The first vintage was 2019, with the grapes brought to Banyuls for vinification. He observed that he has always liked the wines of Aniane: 'they have more finesse than Roussillon'. Emilio joined the business in 2013; he was travelling in Europe and came to Banyuls, working first at Domaine Madeloc for three years.

We tasted an extensive range of wines – what better to do on a very wet morning in Banyuls? Collioure Rosé Côté Mer, from Grenache, Syrah and Carignan, met its objective of a rosé with character, with both fermentation and élevage in oak. It was nicely satisfying. Argile, the *lieu-dit*, based on Grenache Gris, with a little Grenache Blanc, is vinified *à la bourguignonne*, in barrel, and is nicely structured with good acidity. Jean-Emmanuel talked of how his father was a pioneer of this style of wine before the appellation for white Collioure existed; previously it would have been Côte Vermeille. They are convinced that the area is a *terroir à vin blanc*, and others have followed. La Bergerie aux

Abeilles, a blend of Grenache with 10 per cent Carignan, comes from a cooler part of Banyuls and has some fresh, supple fruit. They are looking for cooler sites and giving up the vineyards that really suffer from drought, as they want to avoid heavy, concentrated wines. The red Côté Mer, a blend of Syrah, Grenache and Carignan aged in old foudres, has some weight and pepper. Collioure Côté Montagne comes from north-facing vineyards, including some 80-year-old vines of Grenache and Carignan, as well as Mourvèdre and, unusually, 10 per cent Counoise. It was firm and structured, with youthful fruit. The Counoise comes from the recovered cuttings planted by Dr André Parcé in the late 1970s. No one else has followed their example.

Barlande is mainly Grenache Noir and Carignan, planted together and aged in barrel, and is structured, with some concentration. They would like a *monopole* for this vineyard, and also for a second small vineyard, Col de Perdiguier, but that depends upon the INAO. However, it is Côté Mer and La Bergerie des Abeilles that form the core of their range.

Vin Doux Naturel came next with Voile d'Argile, a Vin de France, with firm acidity and dry, nutty fruit. It had been left outside for two or three years. The nose was more severe than the palate, which had an intriguing, almost honeyed hint on the finish. Rimage Cuvée Thérèse Reig, named after Emmanuel's great-great-grandmother, was muted at 80 grams per litre residual sugar and was redolent of red cherry fruit. Cuvée Léon Parcé, named for his great-grandfather, is from Grenache Noir with some Carignan and is softer, with ripe fruit and less alcohol, after a later bottling and some élevage in old wood. Next came an oxidized Banyuls, Hors d'Age Cuvée Pierre Rapidel, named after a family friend who helped create the local wine growers' syndicat. It was elegantly nutty with a long finish. Our tasting finished with a Rancio Pedro Soler, from Grenache Gris and a little Grenache Blanc, which had spent 15 to 20 years outside in wood, exposed to the elements. It was very concentrated, with a saline nose, firm acidity and an intense nuttiness.

Domaine Clos St Sébastien

Banyuls-sur-Mer

www.clos-saint-sebastien.com

Saint Sébastien is the patron saint of Banyuls, thus providing a fitting name for a wine cellar that was bought in 2008 by Romuald Peronne and his business partner, Jacques Piriou. Their colleague, Pierre Girault, runs the estate in their absence. Pierre worked initially as an archaeologist, specializing in amphora and the commerce of wine, concentrating on Egypt, Greece and Italy. It seemed a very appropriate career change. He observed that the other link between wine and archaeology is the soil. First Pierre took us to see the cellars, in a small hamlet a little way outside Banyuls, along an old smugglers' route near the Musée Maillol. The cellar is well equipped with stainless steel vats, foudres and Stockinger barrels and even some amphorae. They have 20 hectares of vineyards, mostly in Banyuls, in about seven different plots, with varying altitudes and aspects.

Tasting in their seafront shop began with Les Empreintes, a white Collioure from 70 per cent Grenache Gris, with 15 per cent each of Grenache Blanc and Vermentino. It was fresh, with saline tension. 'It is the fingerprint of the terroir,' said Pierre, and absolutely representative of the appellation, combining schist, sea and altitude, to give freshness. Inspiration Minérale is aged in demi-muids for six to nine months and is based on Grenache Gris, from vines that are 60–70 years old, with 10 per cent Grenache Blanc. It was quite firm and oaky, but with obvious ageing potential. Pierre confirmed that white wine is the speciality of the estate, accounting for an unusually high 30–40 per cent of their production of vin sec. Pierre emphasized the freshness of Grenache, providing a counter argument to claims that Grenache only produces heavy wines. The terroir allows for a lot of variations, and he compared Banyuls, with its steep vineyards, to the Valais in Switzerland or the Cinque Terre in Liguria.

Empreintes Rosé, mainly from Grenache, with some Syrah, was lightly rounded with a saline finish. Empreintes Rouge, from Grenache with 10 per cent Carignan, was partly aged in vat and partly in demi-muids, making for some rounded spice with a fresh finish. Inspiration Céleste comes from Grenache with 10 per cent Carignan, grown at 500 metres, with more rounded fruit and a fresh finish, while Inspiration Marine, from 90 per cent Mourvèdre and 10 per cent Grenache, from vineyards near the sea, was concentrated, rich and intense with an elegant finish. Le Clos, from equal parts Grenache, Carignan, Mourvèdre and Syrah, with 12 months in foudres, had silky tannins and fresh fruit, with ageing potential. Sadly, very little of it is made.

Banyuls Blanc comes from Grenache Gris and Grenache Blanc, with just a little Muscat d'Alexandrie. It was rounded and honeyed with a fresh finish. Rimage, a pure Grenache Noir, was redolent of ripe cherries. Banyuls Tradition Inspiration Ardente comes from all three colours of Grenache, aged in *bonbonnes* and barrels for 15 to 20 years without any topping up. There was fruit and liquorice with some tannin and a firm, nutty finish. A 1992 Banyuls *grand cru*, Le Coeur, comes from a selection of bunches; the wine is kept in barrels without any *ouillage*, and the flavours were rich and nutty, with depth and elegance.

Terres des Templiers/Cave Abbé Rous

Banyuls-sur-Mer

www.terresdestempliers.fr

Cave Abbé Rous is the name for the wine trade, and Terres des Templiers is for the consumer. This large cooperative is very well organized to receive visitors, with an enormous tasting area and visitor centre and regular guided visits through the cellars. The director of oenotourism, Clement Pfau, emphasized the importance of tourism to them, with some 60,000 visitors each year. But it is an enormous organization and consequently somewhat impersonal. I was taken round by a friendly guide, Dorothée Collot, who has been with the cooperative for over 20 years. She explained that Terres des Templiers or the Cellier des Templiers, as it was originally called, is made up of several cooperatives, with 750 wine growers farming 1,100 hectares, which represents 55 per cent of the surface of the appellation of Banyuls, 75 per cent of all *grand cru* Banyuls and 70 per cent of all red Collioure.

They use no weedkiller and prune all the vines, except Syrah, as *gobelet*, which must be done before 31 March. There is one organic wine grower with 15 hectares; otherwise everyone else follows *lutte raisonnée*. The technical director, Gregory Cannier, spent six years at Canepa in Chile and has been at Cave Abbé Rous since 2000. A walk through the cellars included a forest of enormous vats, some 100 years old, and the smell was intoxicating. There is also a forest of 600-litre barrels outside; Dorothée talked about the thermal shock the wine receives and said they use a system of spraying to dampen the barrels in order to minimize evaporation.

I tasted a small selection from the extensive range of wine. There was a selection of rosés, an oaky white Schistes de Valbonne, and various red wines. Collioure Les Abeilles was refreshingly unoaked, while Prestige Collioure was very obviously oaky. Various cuvées of Banyuls were named after previous presidents of the cooperatives. Cuvée President Henry Vidal, a pure Grenache kept in small barrels, was quite rich and intense.

I was left with an impression of a rather unwieldly dinosaur and an organization that had once had a more dynamic past.

Domaine La Tour Vieille

Collioure

www.latourvieille.com

I always enjoy my visits to Domaine La Tour Vieille with Vincent Cantié. He has a wicked sense of humour and sprinkles his conversation with perceptive observations. There is an old tower in his vineyards, hence the name of the estate, and altogether he farms 13 hectares in Collioure and Port-Vendres, with the core of the estate in Banyuls. He both rents and owns vineyards. Vincent's grandfather was able to invest in vines at a time when Vin Doux Naturel was profitable. His father gave vineyards to each of his three children and Vincent's first harvest was 1982. His father's main activity was anchovies – like many others, he fished in the summer and worked the vines in the winter.

Collioure Blanc Les Canadells comes mainly from Grenache Gris and Grenache Blanc, with a small amount of Macabeo and Roussanne, and is partly fermented in barrel. The wine has some white blossom, rounded herbal notes and a satisfying mouthfeel. It will age for a few years. Rosé des Roches is a blend of Syrah, Grenache Noir and Grenache Gris and is nicely vinous with a rounded palate. There are three red Collioures. La Pinède comes from Grenache, Mourvèdre and Carignan, with no élevage in wood but a little cool pre-fermentation maceration to retain the aromas and fruit. The wine is immediately appealing, what Vincent called *un vin des copains*, a wine for friends. Puig Ambeille from Mourvèdre, Grenache and Carignan is also kept in vat – Vincent does not like wood – and is firmer and more structured with peppery fruit. Puig Oriol is 70 per cent Syrah with some Grenache and Carignan and a touch of oak, as Syrah needs some oxygen. It was rounded with firm, red fruit.

His Banyuls Blanc is equal parts Grenache Gris and Grenache Blanc. It was rounded and honeyed, with Vincent calling it an amusement. A 2018 Rimage was redolent of ripe fruit and had a touch of cinnamon which was balanced with some acidity and a streak of tannin. A late bottled 2016 Rimage was spicier, with good depth. Banyuls Réserve, given six years of age in *bonbonnes*, barrels and vat had some red fruit and liquorice, balanced with some nutty notes on the finish. We finished with a Banyuls Rancio Solera. The solera was started in 1952 and the wine reminded me of a 40-year-old tawny Port, with an amber colour, rich, nutty fruit and great length on the finish. Vincent very generously sent us home with the bottle.

Domaine du Traginer

Banyuls-sur-Mer

www.traginer.fr

Jean-François Deu explained that his uncle was the last *traginer* of Banyuls. A *traginer* worked with mules, transporting goods from one village to another back when the road system was distinctly limited. His last two mules were called Diego and Mona and they had worked in the vineyards, carrying grapes. Jean-François remembered a harvest from the 1950s employing mules that had come from the mountains and it lasted four to six weeks. His father had made wine, but the domaine did not bottle its own wine until 1982. Although Jean-François has a shop in a side street of Banyuls, his cellar is in the hills outside the town. He has never studied oenology, but learnt everything from his father.

Collioure Rosé, from Grenache, Syrah and little Mourvèdre, was ripe with firm acidity. A 2018 Collioure Blanc comes from Grenache Gris and Blanc, as well as a large amount of Vermentino, which arrived in the region at the beginning of the 2000s. The wine was fragrant and herbal, and very typically Vermentino. The 2016 Collioure Blanc, in contrast, came from Grenache Blanc and Grenache Gris, with a little Tourbat. Jean-François thought he was one of very few vignerons in Banyuls to have the variety. He obtained cuttings from Etienne Montès at Domaine Cazenove, once one of iconic estates of Roussillon, and planted it 15 years ago. It is useful for its acidity. Collioure Blanc from 2015, with Malvoisie as well as Grenache Blanc and Grenache Gris, had stony fruit and showed how well the wine aged. Collioure Rouge is a blend of four grape varieties, the precise blend depending on the vintage, and is rounded and spicy, after ageing in 50-hectolitre foudres. Cuvée d'Octobre spends 14 months in barrel and is nicely structured, and Capitas is predominantly Syrah, with some cedary notes.

Jean-François enjoys experimenting and does what he feels like: 'ça dépend,' he said, with an engaging smile. His first orange wine, Je suis l'Orange, comes mainly from Grenache Gris and Grenache Blanc and is given four weeks maceration and two years of élevage in demi-muids. He observed that it was easier to make than a white wine, when the fermentation can sometimes be hard to finish. The wine had fresh acidity, some tannin, and orange fruit.

Jean-François produces half and half Vin Doux Naturel and vin sec. We began tasting the Vins Doux Naturels with a fresh, spicy Rimage, balanced with a streak of tannin. A *grand cru* had spent eight years in foudres – the minimum is three – and had some firm, liquorice fruit. He tries to use as little sulphur as possible, and has made a *grand cru* without any sulphur, with ripe fruit and good depth. An Ambré was elegantly nutty, with notes of orange marmalade and a hint of honey. Last but not least was a 1991 Hors d'Age. When I asked Jean-François how long it had been in wood, or when it was bottled, he replied, 'I can't remember,' and laughed. It was dry and nutty with a certain structure, an absolutely delicious wine.

Domaine Vial-Magnères

Banyuls-sur-Mer

www.vialmagneres.com

Olivier Sapéras explained that his great-grandfather was the first in the family to make wine, while his father, Bernard, was one of the first to produce a white Banyuls. Olivier's grandfather had a lot of Grenache Blanc in his vineyards, which made the red wines too light in colour if it was blended with Grenache Noir, so his son vinified the white grapes separately, and in 1986 made his first vintage of a white wine, beginning a solera of Banyuls Blanc.

Tasting with Olivier is always fun, as he has an informed and objective view of Banyuls, and on this occasion he taught me a new expression, *cracher derrière la cravate*, to spit behind the tie, or in other words, swallow the wine. But first we had to admire his 'new toys', some eggs made from acrylic, which is easier to wash, and less expensive than clay. He used them for the first time in 2019.

Collioure Blanc Armenn is named after Olivier's great-great-grandfather, who was a lighthouse keeper at Ar-men off the Ile de Sein in Finistère. He originally looked after two local lighthouses at Leucate and Cap Béar, but misbehaved himself with another lighthouse keeper's wife, and the consequence was exile to Brittany. The wine, from Grenache Blanc, Grenache Gris and a little Vermentino, was fresh and floral, with a saline note on the finish. All Olivier's 10 hectares, in 17 plots, are close to the sea, with a beneficial impact on extremes of temperature. Olivier cited an opinion that the local white wines age better than the reds.

His red Collioure Les Espérades, a *lieu-dit*, comes from Grenache with a little Syrah and some Carignan that are fermented together to produce a wine with some pepper and an elegant finish. Olivier's Banyuls Blanc, from Grenache Gris and Grenache Blanc, was fresh with notes of almonds. Olivier would prefer it to be bottled later, but the appellation demands bottling within 18 months. His father kept it for longer. A Rimage was rounded and fresh. Then we moved on, as Olivier put it, to serious things, in other words, the oxidized wines. His *grand cru* spends ten years in barrel, and was nutty with hints of prunes on the nose and palate. It was wonderfully smooth, but with a bite on the finish. Banyuls Ambré Cuvée Bernard Sapéras, named after his father, who began the solera in seven stages in 1986, was redolent of walnuts and hazelnuts, with length and intensity. The solera is topped up with some Banyuls Blanc. In 2019, he could only give the solera a meagre three hectolitres.

Our tasting finished with Altragou, a very old Rancio from Grenache Noir with a little Grenache Gris. The meaning of *al tragou* is intriguing, but difficult to explain; a tiny stream of wine between the mouth and the bottle. The wine was *tuilé* in colour, with firm fruit, dry prunes, incisive acidity, great length and a long, saline finish. I found it irresistible.

It is difficult to make money in Roussillon as a wine grower and Olivier does not intend to encourage his teenage children to go into the wine trade, even though he clearly loves making wine. It was the last visit of a wonderful week in Banyuls and I was left wondering just why these wines are so under-appreciated.

APPENDIX I: VINTAGES, 2000–2020

There is no history or tradition in Roussillon for ageing vin sec, partly as its production is relatively recent. There are exceptions of course, with Frédérique Vaquer coming to mind, and also Olivier Pithon. Vin Doux Naturel is another story, but it is usually aged in a cellar, and kept in barrel until bottling, and may well be a blend, an Hors d'Age without a vintage date. You may occasionally encounter a wine grower with a stash of old bottles, such as Domaine de Rancy and Georges Puig. Consider too the cooperative in Baixas and L'Etoile in Banyuls where Bruno Cazes has worked to save the fabulous stocks of old Banyuls.

The other problem about giving vintage details is that the microclimates of Roussillon complicate matters. The climate is extraordinarily varied, with enormous differences between say, Espira-de-l'Agly and Caudiès-de-Fenouillet at either ends of the Agly Valley. So what follows are some generalizations as well as individual observations, substantiated by the annual vintage reports from the CIVR. It could also be said that the vintage differences are details; there has never been a year when the grapes have not ripened sufficiently. The key questions are rainfall and drought, occasionally hail, but rarely mildew. The grapes need to be riper for Vin Doux Naturel than for vin sec.

2020: Very unusually, mildew is the dominant theme of 2020. While London enjoyed an early summer during the Covid-19 spring lockdown, Roussillon experienced days of damp weather, resulting in attacks of mildew, with very little of the wind that would normally have dried the vegetation. Over four days at flowering 200 millimetres of rain fell. The summer was warm, resulting in a very early harvest, one of the earliest ever, not just of Muscat but other varieties too. Sébastien Danjou started picking on 7 August but did not finish until late September, and at Coume del Mas harvest began on 11 August. The very wet spring weather helped keep the local Grenache vines, which are well used to summer drought, in good shape. Daniel Laffite found the year difficult because of the mildew, but admitted from his early impressions that it had turned out better, even much better, than he had initially expected. Certainly, the few vat samples that I tasted in late September promised well, and most wine growers were cheerful about their quality.

2019: Very good quality, but low quantity, with approximately 600,000 hectolitres. July was hot and dry, while some rain in September brought a bit of beneficial freshness to the wines. Olivier Sapéras talked about sunburn and Quentin Modat mentioned a hot June, but two beneficial storms later in the summer. Others put 2019 down as a drought year.

2018: A wet spring replenished water supplies after a dry winter and this was combined with the beneficial effects of the tramontane wind. A very hot August was balanced by regular rain so there was no water stress and the quantity was normal. The harvest lasted from the end of July to the beginning of October for vin sec, and to the middle of October for Vin Doux Naturel.

Amongst growers, opinions about the year vary. Jean-Philippe Padié described the year as schizophrenic, with the very wet spring provoking mildew, followed by a hot and dry July and August. He began his harvest with Mourvèdre, which he usually picks last. Domaine Fontanel lost crop to mildew; Quentin Modat also mentioned the mildew, and did not find the wines as balanced as in 2019. Château de Corneilla saw the summer as one of drought. Domaine des Soulanes suffered hail in early June which affected 9 hectares. For Vincent Cantié, 2018 produced his largest crop since his first harvest in 1982, totalling 500 hectolitres (by contrast 2019 was his smallest ever at just 320 hectolitres; his average from his 13 hectares is 440 hectolitres). Domaine la Toupie said the year produced elegant wines and Frédérique Vaquer said the wines were delicate.

2017: A wet winter and spring but the spring was also warm, so bud break and flowering were early. However, unusually, 500 hectares were affected by frost damage, mainly vineyards on lower-lying land, producing IGPs. Domaine Fontanel lost two hectares to hail and Château de Corneilla also suffered from hail. Drought was an issue at Domaine Piquemal. It was a more concentrated vintage, with very healthy grapes and more tannins. Gérard Gauby started the harvest on 31 July and finished on 16 September; Roc des Anges quickly followed him on 3 August. Gérard lamented the obvious effects of global warming.

2016: There was no rain and yields were very small, with small, concentrated berries, after a dry spring and a hot summer. Old vines fared better than young vines. Joseph Paillé talked about making his first wines from skins and pips, so dry it was, with an annual rainfall of just 125 millimetres. Marie-Pierre Piquemal also talked about the drought, which lasted from June to September, and like Joseph noted that there were grapes but no juice. Harvest began in mid-August. Jean-Philippe Padié said the heat had made for dense wines, while Jérôme Collas at Domaine la Toupie found elegance, and others observed that despite the drought the grapes had retained acidity.

2015: Flowering was early and the harvest began with Muscat à Petits Grains on 7 August, with very healthy grapes. Charles Perez started picking earlier than usual on 12 August. For Bruno Ribière, it was one of the best vintages for quality, with elegance and balance as well as ageing potential. Jean-Roger Calvet enthused about the quality of the vintage but Jérôme Collas found the wines too powerful: 2016 and 2018 were more elegant for him. Domaine Vaquer's L'Exception was savoury and elegantly balanced, drinking beautifully in the winter of 2020, as was Domaine Gardiés's Clos des Vignes.

2014: The harvest began on 15 August and continued until early October, with very healthy grapes. The nights were cool and some storms provided beneficial rain, making for balance, fruit and good acidity in the wines, allowing for greater ageing potential. Charles Perez finds the wines elegant, and similar to those of 2012. Couillades d'en Paillol from Domaine Paul Meunier Centernach was ripe with elegant tannins.

2013: A summer of cool nights and warm days. Some rain in early July and also September had a beneficial effect, as did the tramontane wind, which kept everything healthy. Quality is very good, with some concentrated wines, and the vintage is comparable to 2007 or 2011. It has been described as a year of colour.

2012: A cold, wet winter was followed by a warm spring and a hot, dry summer, with an August heatwave. The tramontane wind kept everything healthy, but yields were low, the lowest for 40 years, with a lot of berries dropping in the heat and a total production of 700,000 hectolitres. The wines have aroma and concentration and some growers have described it as a Grenache year. There was also hail in Maury and the Fenouillèdes, with a storm in early August. Jean-Roger Calvet only made his entry-level wine.

2011: A wet start to the year, particularly January and March, made for good water reserves. April and May were warm, and a cooler than usual July was followed by a hot August. There was some hail in Les Aspres. Generally, the tramontane wind kept everything healthy, as it so often does. The harvest took place in good conditions. For Frédérique Vaquer, the quality was very good and wines have acidity, freshness and tension. Carignan performed particularly well.

2010: A cool, dry winter was followed by a warmer than average April. Temperatures in July were also above average and some rain fell in August. Conditions at flowering were good but the final crop was 30 per cent below average, with small berries. For Eric Monné at Clot de l'Oum it was 'une belle année', without too much alcohol. Charles Perez at Mas Bécha described it as a fresher year, and Olivier Pithon said it was a very fresh, cool year, with a small production. The wines have good acidity and are riper than those from 2011. Clos des Fées Vieilles Vignes was ripe, spicy and mouth filling with a harmonious finish.

2009: The harvest began earlier than usual on 17 August. The tramontane wind had a beneficial effect on the health of the grapes, and resulted in some concentrated wines. Both Olivier Pithon and Charles Perez talked of a hot year and even drought.

2008: A very good year. There was a summer drought, with hot, dry conditions and very little wind, making for small berries. The crop was small and healthy, and a little rain on 11 September had a beneficial effect on the later-harvested grapes. A comparison of Domaine de la

Rectorie Côté Mer and Côté Montagne was fascinating. Mer had firm tannins, with structure and peppery fruit while Montagne was denser and more tannic, but still youthful in the summer of 2020. Domaine Gauby's Vieilles Vignes was drinking beautifully as I was putting the finishing touches to this book in the autumn of 2020.

2007: The harvest began early, on 2 August, with Muscat à Petits Grains. The grapes were healthy, the tramontane helped the concentration, and the volume was the lowest hitherto at just over 900,000 hectolitres, with an average of 30 hectolitres per hectare. Quality was good, even very good; Jean-Roger Calvet said it was a very good vintage, with everything ripe, but not overripe. Somebody observed that it ended the curse of the difficult years ending in 7, namely 1967, 1977, 1987 and 1997.

2006: A cold winter with good rainfall preceded a hot dry summer; the tramontane blew in August. Some beneficial rain fell around 15 August and again on 15 September. Small, healthy berries made for a small volume and concentrated flavours. Eric Monné described the vintage as 'une année ingrate', an awkward, complicated year. He did not like the wines early on so forgot about them for a number of years before rediscovering and selling them. Frédérique Vaquer, on the other hand, said the wines had freshness and very good balance, and were destined for a long life, but are overshadowed by 2005 and overlooked. She thinks them similar to those of 2016. Domaine Gauby's Vieilles Vignes enjoyed in the autumn of 2020 was ageing beautifully, with perfumed fruit and a streak of tannin. It had more concentration than the 2008, while his Muntada was even richer.

2005: A cold winter with good rainfall was followed by a cool spring. July and early August were hot, dry and windy. Some rain in early September had a beneficial effect as fine weather followed and the harvest took place in good conditions. The small crop has fulfilled its initial ageing potential. Frédérique Vaquer thinks it a splendid year, similar to 2015, and there is no doubt that her 2005 was drinking beautifully in the summer of 2020.

2004: A year of rain, although not as much as in 2002. The harvest was complicated and required a lot of triage. Quite a large crop, with over 1,360,000 hectolitres.

2003: The year of the heatwave, but the spring was wet and the temperatures were not really excessively hot for Roussillon – it is all relative. A little rain in the middle of August aided ripening. September was sunny. Acidity was low and the grapes had relatively little juice. Maury was affected by hail. Olivier Pithon remembers a very hot July and August, but finds that the wines are ageing well.

2002: This is a year that tested the wine grower. The spring was mild and wet, and the summer initially dry, but August was cool and wet. The tramontane in September helped to dry things out, but with so much rain, a lot of sorting was essential at harvest. The yield was low and the grapes were not very healthy.

2001: The summer was hot and dry, with no mid-August storm. The harvest was early, with healthy grapes, making for good quality wines. Bruno Ribière said it had been a year of climatic accidents (but didn't elaborate further), but Frédérique Vaquer remembered what she called 'une jolie chaleur', some nice warmth.

2000: A relatively dry winter was followed by a wet spring. The summer was dry and windy, and conditions were hot and dry during the harvest, making for a good volume of healthy grapes, with some ageing potential.

APPENDIX II: PRODUCTION FIGURES, AREA AND VOLUME FOR THE 2019 HARVEST[4]

	Area (ha)	Volume (hl)
Banyuls + Grand cru	882	12,512
Maury	188	3,071
Muscat de Rivesaltes	3,248	58,733
Rivesaltes	1,548	36,421
Total Vin Doux Naturel	5,866	110,737
Pyrénées-Orientales	5,636	106,565
Aude	229	4,172
Collioure Blanc	97	2,400
Collioure Rosé	109	2,650
Collioure Rouge	247	6,939
Total Collioure	453	11,989
Côtes du Roussillon Blanc	425	11,691
Côtes du Roussillon Rosé	1,714	63,064
Côtes du Roussillon Rouge	2,312	74,200
Total Côtes du Roussillon	4,451	148,955

4 Figures have been rounded to the nearest whole number.

Maury Sec	273	4,435
Côtes du Roussillon Villages	1,756	50,075
Caramany	258	7,305
Latour-de-France	189	4,916
Les Aspres	177	5,535
Lesquerde	70	2,083
Tautavel	337	10,431
Languedoc Blanc	—	—
Languedoc Rosé	103	4,162
Languedoc Rouge	16	585
Total Languedoc	119	4,747
AOP Sec	7,983	250,471
AOP Roussillon	7,864	245,724
AOP Languedoc[5]	119	4,747
Côte Vermeille	14	303
Côtes Catalanes Blanc	724	31,796
Côtes Catalanes Rosé	956	41,697
Côtes Catalanes Rouge	2,017	74,107
Total Côtes Catalanes	3,697	147,600
Pays d'Oc Blanc	647	33,305
Pays d'Oc Rosé	266	15,905
Pays d'Oc Rouge	518	25,191
Total Pays d'Oc	1,431	74,401
Total IGP	5,142	222,304
Vin de France	500	18,800

Average yield 31.47 hectolitres per hectare

5 To clarify a possible confusion, Languedoc AOP is included in this chart as Côtes du Roussillon can be declassified into Languedoc if it is to be blended with wine from the Languedoc. There are Languedoc producers who buy wine in Roussillon in a négociant capacity.

APPENDIX III: AREA PLANTED BY VARIETY, 2019

Red grape varieties	Area in hectares
Grenache Noir	5,942
Syrah	4,443
Carignan	3,250
Mourvèdre	922
Lledoner Pelut	160
Cinsault	75
Merlot	509
Cabernet Sauvignon	412
Cabernet Franc	73
Marselan	184
Chenanson	24
Total	**15,994**

White grape varieties	
Muscat à Petits Grains	2,491
Muscat d'Alexandrie	2,047
Macabeo	1,551
Grenache Blanc	1,274
Grenache Gris	964
Tourbat (Malvoisie du Roussillon)	31
Chardonnay	484

Viognier	119
Sauvignon Blanc	74
Vermentino	142
Roussanne	79
Marsanne	32
Total white	**9,288**
Sundries in both colours	280
Total	**25,562**

APPENDIX IV: A HISTORY OF THE VINEYARD IN HECTARES[6]

Year	Hectares
1741:	9,000
1788:	26,000
1820:	40,000
1850:	50,000
1870:	55,000
1876:	58,000
1880:	73,000
1882:	76,000
1891:	42,000
1911:	61,000
1935:	72,000
1960:	69,000
1979:	58,000
1990:	44,000
1995:	40,000
2000:	38,000
2005:	30,000
2008:	29,000
2009:	24,000
2010:	23,900
2014:	21,800
2016:	21,400
2017:	20,900
2018:	20,700
2019:	19,900

6 Figures are rounded to nearest hundred.

APPENDIX V: PRODUCTION FIGURES

Year	Volume (hl)			Area (ha)		
	Vins Secs	Vin Doux	Total	Vins Secs	Vins Doux	Total
2000	1,265,263	364,271	1,629,534	21,021	14,396	35,417
2010	555,800	197,655	753,455	15,409	10,143	25,552
2015	683,890	139,145	823,035	15,192	6,674	21,866
2018	623,702	131,782	755,484	14,652	6,004	20,656
2019	515,756	106,565	622,321	14,039	5,635	19,674

Muscat de Rivesaltes[7]

Year	Volume (hl)	Area (ha)
2000	155,515	4,809
2010	117,753	4,756
2019	54,940	3,033

Rivesaltes

Year	Volume (hl)	Area (ha)
2000	131,958	6,782
2010	61,996	4,233
2020	36,042	1,533

7 The difference between this figure for Muscat and the figure in Appendix II is explained by Muscat Sec, or Muscat in blends of IGP, and also the possibility of including some Muscat in Rivesaltes Ambré.

APPENDIX VI: THE GROWTH OF ORGANIC VITICULTURE IN ROUSSILLON

Year	Vineyard area (ha)	Organic vineyards (ha)	Organic vineyards (%)
2011	25,252	2,940	12%
2012	23,866	3,054	13%
2013	23,496	2,937	13%
2014	23,640	3,153	13%
2015	22,533	3,557	16%
2016	25,721	3,647	14%
2017	21,438	3,827	18%
2018	21,084	4,263	20%
2019	19,905	4,976	25%

GLOSSARY

Agrément. Official acceptance of a wine for its appellation.

Are. A historical unit of measurement, equivalent to 100 square metres.

Arrachage. Pulling up of vines; in the past growers were paid a subsidy to do so with *primes d'arrachage*.

Assemblage. Blend.

Ban des vendanges. Official harvest date.

Barrique. Barrel of 225 litres.

Bâtisse. A large building or other edifice.

Bâtonnage. Lees stirring.

Bonbonne (not bonbon). Demijohn or large glass jar used to store wine outside to enhance the *rancio* character.

Bordelais. Adjective relating to Bordeaux, often used in the context of barrels, or a person.

Cadastre. A legal document describing land-holdings.

Cahier des charges. The document that lays down the regulations of the appellation and what the wine grower may or may not do.

Calcaire. Limestone.

Capitelle. Little drystone building in a vineyard, that looks like an igloo.

Carbonic maceration. A method of fermentation, whereby whole bunches are put into a vat filled with carbon dioxide and left to ferment. The fermentation process begins inside the grapes, and contrary to a usual fermentation, the pressed juice is better than the free-run juice. It is probably less important in Roussillon than it used to be.

Caviste. Wine shop.

Cépage. Grape variety.

Cépage améliorateur. Describes grape varieties like Syrah, Grenache Noir and Mourvèdre, which were seen to improve the once despised Carignan and Aramon.

CIVR. Conseil Interprofessionel des Vins du Roussillon. The professional body of the Roussillon wine trade.

Compagne (f) Compagnon (m). Partner with an emotional attachment, but not necessarily a working relationship.

Complanté. Grape varieties planted together in a vineyard, making for a field blend.

Confit. Literally, candied, but describes a wine that is rich, concentrated and almost sweet.

Coulure. A disorder of the vine which can occur if flowering takes place in unsatisfactory climatic conditions, so that the berries fail to develop, adversely affecting the yield.

Court noué. Fan leaf, a virus of the vine spread through the soil. Old vines are particularly affected.

Cru. Literally a growth, and part of the French wine hierarchy with *premier cru* and *grand cru.*

Cuvaison. Time in vat with grape skins in contact with the juice.

Cuvée. Almost a synonym for a wine; a wine grower will produce several different cuvées or wines.

Débourbage. The process of clarifying the juice of white wine before fermentation begins, usually by chilling for several hours.

Délestage. Rack and return is the best English translation. The juice is run out of the vat, so the cap of grape skins sinks to the bottom and then, when the juice is returned to the vat, the cap rises up again through the juice.

Demi-muid. A barrel, usually of about 500 to 600 litres.

Eboulis calcaires. A mass of fallen rock, limestone.

Ebourgeonnage. Debudding.

Egrappé. Destalked.

Egrappoir. The machine that destalks.

Elevage. Literally the rearing, 'educating', or ageing of wine.

En echalas. A vine planted with just one supporting post and without any wires.

Enherbement. Ground cover, leaving grass, etc., in the vineyard.

En vrac. Wine in bulk.

Esca. Vine trunk disease.

Fermage. Term for renting a vineyard.

Field blend. A mix of grape varieties in a vineyard, usually an old vineyard. *Complanté* in French.

Finesse. Tasting term that implies elegance.

Flavescence dorée. One English translation is 'grapevine yellows'. A disease of the vine, spread by leafhoppers, which kills young vines and severely affects the yield of older vines. A potentially serious problem.

Flor. See *voile*.

Foudre. Large barrel, anything from 5 to 500 hectolitres.

Foulage. Breaking the grape skins to release their juice.

Fût de chêne. Oak barrel, usually small.

Galets roulés. Large, round pebbles, most typical of the vineyards of Châteauneuf-du-Pape, but also found in Roussillon.

Garrigue. Mediterranean scrubland, with the typical vegetation of laurel, thyme, juniper, cistus and other aromatic plants.

Gobelet. Bush vines.

Gouleyant. Drinkable.

Gourmand. Literally greedy, but when describing a wine, implies rich, warm appealing flavours.

Grand cru. Literally a great growth, above *premier cru* in the pecking order.

Grenat. Relatively recent term for the youngest and freshest Rivesaltes and Maury, requiring just eight months of ageing.

Gris. A term used to describe a grape variety that has a pale pink skin when ripe, such as Grenache Gris as opposed to Grenache Noir or Grenache Blanc.

HVE. Haute Valeur Environnementale. Farming method that places an emphasis on biodiversity.

IGP. Indication Géographique Protégée, the term which has replaced *vin de pays*.

INAO. Institute National des Appellations d'Origine, the government body that controls all the details of an appellation.

Intercep. Piece of vineyard equipment that enables you to weed mechanically between the vines within the row.

Jarre. Amphora.

Labelle. Term used for the tasting session which ensures that a wine is of acceptable quality for an appellation.

Lieu-dit. Small plot of land on a map, literally 'place named'.

Lutte raisonnée. Sustainable viticulture; the grower thinks before spraying, rather than spraying irrespective of climatic conditions.

Mairie. Town hall, even in a small village.

Malolactic fermentation. The transformation of malic acid (such as that found in apples) into softer lactic acid (as found in milk). Usually occurs immediately after the alcoholic fermentation.

Mas. A farmhouse, especially used in Provence.

Massal selection. Propagating by selecting vine cuttings from a vineyard, rather than buying clones from the nursery.

Mistelle. Grape juice, with the addition of alcohol so that it does not ferment. Often used as the base of sweet aperitifs.

Moelleux. Lightly sweet.

Monocépage. Single grape variety in a wine.

Monopole. A single named vineyard.

Mouillage. Dampening the cap of grape skins.

Mutage. The stopping of the alcoholic fermentation by the addition of grape spirit, for the production of Vin Doux Naturel. When the wine is *muté sur grains*, the alcohol is added before the juice has been run off the skins, and when *muté sur jus*, the alcohol is added just to the juice. For the sake of simplicity I have anglicized the term to 'muted'.

Négoce. Collective noun for several négociant companies.

Négociants. Merchants who buy grapes, juice or wine in varying stages of preparation and then make their own blends.

Occitan. The *langue d'oc*, the original language of the Languedoc, also spoken in other Mediterranean areas and today enjoying something of a revival.

Ouillage. Topping up of barrels, when wine has been lost naturally by evaporation.

Pain d'épice. A French speciality, a cross between cake and bread, flavoured with spices.

Palissé/palissage. Describes vines that are trained on wires.

Passerillé. Raisined grapes, dried on the vine in the autumn sunshine.

Pêche de vigne. A type of small peach with red flesh, often found in the vineyards, hence its name.

Pétillant. Lightly sparkling.

Pét nat. An abbreviation of pétillant naturel, i.e. naturally sparkling.

Pièce. A Burgundian barrel of 228 litres.

Pigeage. Pushing down the cap of grape skins.

Pinotte. When a wine takes on notes of maturing Pinot Noir. A term more commonly used in relation to Beaujolais, but since Grenache Noir is sometimes called the Pinot Noir of the south it is also relevant here.

Premier cru. A first growth, and usually below *grand cru* in the hierarchy.

Prestateur. Literally someone who provides a service.

Rancio. Wine aged in barrel, so that it oxidizes gently and deliciously; usually fortified, making a delicious dessert wine. An old tradition.

Rapport qualité prix. Price quality ratio.

Remontage. Pumping over the cap of grape skins in order to extract colour and flavour.

Rimage. Term for the youngest and freshest style of Banyuls, requiring ageing for just eight months.

Saignée. Term used for making rosé, literally means 'bleeding' the vat, and running off juice.

Salivant. Makes you salivate, mouth-watering.

Sélection parcellaire. Site or plot selection.

Stage. Work experience, or a short apprenticeship.

Sur grains. The addition of alcohol on the grape skins.

Sur jus. The addition of alcohol on the fermenting juice.

Syndicat. Union, as in the growers' unions that run the appellations.

Terroir. A French term impossible to translate directly into English. It includes soil as well as aspect and altitude, and encompasses the overall environment of the vine. It is also used to describe an area within an appellation, e.g. Les Aspres, before it became part of Côtes du Roussillon Villages, was considered a terroir of Côtes du Roussillon.

Tonneau. A large barrel, usually 900 litres or 4 barriques.

Tonnelier. Cooper.

Tronconic vat. A large, tapered vat that may be in wood, stainless steel or concrete and which is used for fermentation or élevage.

VDQS. Vin Delimité de Qualité Supérieure, the category beneath *appellation contrôlée* and now virtually defunct.

Vers de la grappe. Larvae from two insects, cochylis and eudemis, that can do considerable damage to vines.

Vigneron/ne. Wine grower; they make their own wine as well as growing grapes.

Vin de garde. A wine intended for ageing.

Vinification intégrale. A recently developed technique entailing the fermentation of red grapes in a small oak barrel, from which the top is removed so that it is possible to put the grapes in the barrel.

Viticulteur. Grower of grapes; does not make wine.

Voile. A veil of yeast, often found in Rancio Sec and always in fino sherry, where it is known as *flor*.

BIBLIOGRAPHY

Bray, Richard W. H., *Salt & Old Vines*, Unbound, 2014

Camo, Louis, et. al., *Les Vins du Roussillon*, Editions Montalba, 1980

Deyrieux, André, *A la rencontre des cépages modestes et oubliés*, Paris: Dunod, 2016

Dion, Roger, *Histoire de la Vigne et du Vin en France des Origines au XIXème siècle*, Paris: Flammarion, 1959

George, Rosemary, *French Country Wines*, London: Faber & Faber, 1990

George, Rosemary, *The wines of the South of France, From Banyuls to Bellet*, London: Faber & Faber, 2001

Gorley, Peter, *The Wines and Winemakers of Languedoc-Roussillon*, Hamilton John Publishing, 2015

Guyot, Jules, *Etudes des Vignobles de France*, Paris, 1868

Healey, Jonathan, *The Wines of Roussillon*, Canet: Editions Trabucaire, 2002

Healy, Maurice, *Stay Me with Flagons*, London: M. Joseph Ltd, 1949

Jefford, Andrew, *The New France*, London: Mitchell Beazley, 2002

Johnson, Hugh, *The World Atlas of Wine* (1st edn), London: Mitchell Beazley, 1971

Johnson, Hugh and Jancis Robinson, *The World Atlas of Wine* (8th edn), London: Mitchell Beazley 2019

Jullien, André, *Topographie des Tous les Vignobles Connus*, Paris, 1866

Lachiver, Marcel, *Vins, Vignes et Vignerons: Histoire du vignoble français*, Lille: Fayard, 1988

Legeron, Isabelle, *Natural Wine*, London: CICO Books, 2014

Monferran, Florence, *Les Breuvages d'Héracles*, Toulouse: Privat, 2020

Pomerol, Charles, (ed.), *Terroirs et Vins de France*, Paris: Total-Edition-Presse, 1984

Pottier, Alain, *Les Rancios Secs du Roussillon*, Canet: Editions Trabucaire, 2016

Rendu, Victor, *Ampélographie Française*, Paris, 1857

Robinson, Jancis, Julia Harding and José Vouillamoz, *Wine Grapes*, London: Penguin, 2012

Robinson, Jancis and Julia Harding, *The Oxford Companion to Wine* (4th edn), Oxford: Oxford University Press, 2015

Rosenstein, Jean-Marie, *BYRRH, la Saga des Violet*, Nîmes: Mondial Livre, 2018

Shand, Morton J., *A Book of French Wines*, London: Penguin, 1964

Sichel, Allan, *The Penguin Book of Wines*, London: Penguin, 1968

Strang, Paul, *Languedoc Roussillon: The Wines and Winemakers*, London: Jean and Paul Strang Partnership, 2017

Waldin, Monty, *Biodynamic Wine*, Oxford: Infinite Ideas, 2016

Wilson, James, *Terroir*, London: Mitchell Beazley, 1998

ACKNOWLEDGEMENTS

Books about wine are about vineyards and bottles, but above all they are about people, so I would like to thank all the wine growers of Roussillon who contributed to this book by taking the time to show me their vineyards, in some wonderful scenery, and their cellars, of varying sizes and degrees of modernity. They opened bottles and answered questions, sharing their enthusiasm and passion for the wines of Roussillon, but also their disappointments, aspirations and challenges. Without them, there would be no book. André Dominé, a fellow wine writer, who has lived in Roussillon for many years, offered further insights.

The book could not have happened without the invaluable help of Eric Aracil, the Directeur Adjoint et Responsable Service Export of the CIVR, to give him his full title. Eric has worked with the wines of Roussillon for nearly 25 years, mainly with the CIVR, and what he does not know about the wines of Roussillon is simply not worth knowing. Along with his colleagues Hélène Losada and Charles Husson he has patiently answered questions and supplied statistics and supplementary information. Eric and Hélène also scoured my text for errors of fact and French. The CIVR arranged accommodation, in *chambres d'hôtes* in Maury and Calce, and a welcoming little gîte at Domaine des Soulanes, right in the heart of the vineyards. My good friends Liz Morcom MW and her husband Roger Baggallay also offered me the use of the studio flat in their lovely house in the hills above Ille-sur-Têt. It was a wonderful tranquil haven after a day's intensive wine tasting.

Photographs that capture the dramatic scenery and character of Roussillon were supplied by the CIVR, Domaine de la Rectorie and Mas Amiel.

Richard Burton at Infinite Ideas enabled me to explore in greater depth another of my favourite regions of France. And thanks also to the editorial team, including Rebecca Clare for her beady-eyed perusal of my text, the proofreader, Sue Morony, and Catherine Hall for creating the index. Maps were created by Darren Lingard from originals supplied by the CIVR.

Last but certainly not least, my husband, Christopher Galleymore deserves my grateful thanks for living with the trials and tribulations of a gestating book. As always, he has been immensely supportive and has often shared the research, and acted as my chauffeur. Fortunately, he enjoyed discovering the wines of Roussillon as much I did.

London, February 2021

INDEX

More from The Classic Wine Library

The Classic Wine Library series is a premium source of information for students of wine, sommeliers and others who work in the wine industry, but can easily be enjoyed by anybody with an enthusiasm for wine. All authors are expert in their subject, with years of experience in the wine industry, and many are Masters of Wine. The series is curated by an editorial board made up of Sarah Jane Evans MW, Richard Mayson and James Tidwell MS. Explore more of France with the titles below.

Wines of the Languedoc

For those who like individual wines that express the place where they were grown, the Languedoc is a perfect hunting ground. Rosemary George MW presents the most interesting producers and unpacks the region's wines for wine professionals and consumers alike. *"The most important book written on the Languedoc so far."* – Tamlyn Currin, jancisrobinson.com

Wines of the Rhône

This guide to one of the great French wine regions covers all the appellations of the Rhône, featuring interviews with some of the most respected winemakers of the region, and tackles the issues facing the Rhône's wines with clarity and authority. An ideal introduction for those new to the Rhône, while providing fresh insights for long-time admirers of the wines.

The wines of Chablis and the Grand Auxerrois

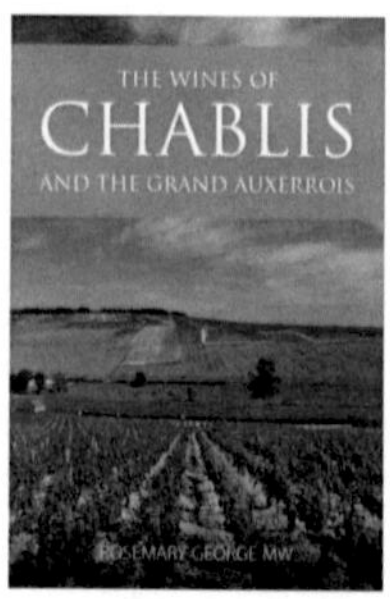

Features the history, vineyards, *crus* and viticultural methods of Burgundy's most northerly wine-growing region, profiling the producers who make Chablis in the twenty-first century and assessing what the future holds for these historic wines. The book also explores the wines and producers of the neighbouring wine region, the Grand Auxerrois.

913141813